T0192830

Ultra-low Voltage Low Power Active-RC Filters and Amplifiers for Low Energy RF Receivers

Lucas Compassi Severo • Wilhelmus Adrianus
Maria Van Noije

Ultra-low Voltage Low Power Active-RC Filters and Amplifiers for Low Energy RF Receivers

 Springer

Lucas Compassi Severo
Federal University of Pampa
Alegrete, Rio Grande do Sul, Brazil

Wilhelmus Adrianus Maria Van Noije
Universidade de São Paulo
São Paulo, São Paulo, Brazil

ISBN 978-3-030-90105-9   ISBN 978-3-030-90103-5   (eBook)
https://doi.org/10.1007/978-3-030-90103-5

This Springer imprint is published by the registered company Springer Nature Switzerland AG
The registered company address is: Gewerbestrasse 11, 6330 Cham, Switzerland

For my son Thomas and my wife Tanísia

Lucas Compassi-Severo

For my wife Ana Franklin

Wilhelmus Van Noije

Preface

The internet of Things (IoT) and it derivations are changing the environment we live in by increasing the number of electronics devices connected to the cloud and to each other. The IoT is very important for achieving an improved quality of and also for increasing industry productivity. In spite of that, the massive number of connected devices increased the demand for low-energy circuits able to operate using energy harvesting or small batteries.

The RF transceivers are the main power-hungry part of an IoT device. In these circuits, the high-frequency parts present the most power dissipation. However, modern communication standards have reduced the RF circuit's requirements, resulting in substantial power optimization. As a consequence, the power dissipated by the intermediary frequencies and base-band circuits parts become more expressive on the receiver's total power dissipation. This has motivated the development of modern circuits topologies and design methodologies for active filters and programmable gain amplifiers (PGA) able to operate with ultra-low voltage (ULV) and to present ultra-low power (ULP) dissipation.

In this book, the ULP dissipation is obtained by using very efficient single-stage inverter-based operational transconductance amplifiers (OTA) and proper forward bulk biasing to reduce the sensitivity to process voltage and temperature (PVT) variations. The low voltage gain and the resistive load effects on the single-stage OTA are completely compensated by using a PVT robust negative transconductor connected at the OTA inputs. The dynamic range is increased by using fully differential topologies and common-mode feedback to improve the common-mode and power supply rejection rates. The operation at the ULV range is reached by using only two-stacked transistors in all the circuit implementations and bulk forward bias in some transistors to reduce the threshold voltage and to increase the channel inversion level.

An operation point simulation-based tool and some design methodologies are also presented in this book to design the ULV circuits. The presented circuits were used to design a third-order active-RC complex band-pass filter (CxBPF), a programmable gain amplifier (PGA), and a Tow-Thomas biquad, with integrated programmable gain capability, suitable for BLE RF receivers. The PGA implementation uses a new programmable input negative transconductor to obtain the optimal closed-loop amplifier compensation in all the gain modes. The circuits were designed to operate at the power supply voltage of 0.4 V and are prototyped in 180 nm and 130 nm low-cost CMOS and BiCMOS process, respectively.

The experimental and post-layout simulation results shown in this book have demonstrated the proper ULV operation at 0.4 V, the ultra-low power dissipation down to 10.9 μW/pole, and the best figure-of-merit (FoM) among the state-of-the-art active-filters and amplifiers from the literature.

Alegrete, Brazil   Lucas Compassi Severo

São Paulo, Brazil   Wilhelmus Adrianus Maria Van Noije

Acknowledgments

We would like to thank all the students, colleagues, and friends from the University of São Paulo and the Federal University of Pampa who have collaborated to the development of the researches and integrated circuit designs presented in this book.

We also would like to thank the partial financial support from the Brazilian National Council for Scientific and Technological Development—CNPq (Grant number 141079/2015-0), the Foundation of the State of Rio Grande do Sul—FAPERGS (Process Number 19/2551-0001315-9), and the Polytechnic School of Engineering at the University of São Paulo.

We would like to thank the staff at Springer, in particular Charles Glaser and Pradheepa Vijay, for their help and support.

Contents

Acronyms

AC	Alternating Current
BiCMOS	Bipolar and Complementary Metal-Oxide-Semiconductor
BLE	Bluetooth Low Energy
BPF	Band-Pass Filter
CAD	Computer-Aided Design
CM	Common-Mode
CMFB	Common-Mode Feedback
CMOS	Complementary Metal-Oxide-Semiconductor
CMRR	Common-Mode Rejection Ratio
CxBPF	Complex Band-Pass Filter
DC	Direct Current
DDA	Differential Difference Amplifier
DFF	D-Type Flip-Flop
DIBL	Drain-Induced Barrier Lowering
DL	Diffusion Region Length
DM	Differential-Mode
DR	Dynamic Range
FD	Fully Differential
FF	N-Fast and P-Fast Process Corner Parameters Set
FoM	Figure of Merit
FS	N-Fast and P-Slow Process Corner Parameters Set
GBW	Gain-Bandwidth Product
GUI	Graphical User Interface
IF	Intermediate Frequency
IIP_3	Input Third-Order Intermodulation Intercept-Point
IM_3	Third-Order Intermodulation Product
IRN	Input-Referred Noise
IRR	Image Rejection Ratio
LNA	Low Noise Amplifier
LOD	Length of Diffusion Effect
LPF	Low-Pass Filter

MI	Moderated Inversion
MiM	Metal-insulator-Metal
MOS	Metal-Oxide-Semiconductor
NF	Noise Figure
NIC	Negative Input Converter
NMOS	Metal-Oxide-Semiconductor type N
OPAMP	Operational Amplifier
OPD	Operation-Point Driven
OTA	Operational Transconductance Amplifier
PDK	Process Design Kit
PGA	Programmable Gain Amplifier
PMOS	Metal-Oxide-Semiconductor type P
PSRR	Power Supply Rejection Ratio
PVT	Process Voltage and Temperature
RF	Radio Frequency
RSCE	Reverse Short-Channel Effect
SI	Strong Inversion
SF	N-Slow and P-Fast Process Corner Parameters Set
SFDR	Spurious-Free Dynamic Range
SS	N-Slow and P-Slow Process Corner Parameters Set
STI	Shallow Trench Isolation
THD	Total Harmonic Distortion
TIA	Transimpedance Amplifier
ULP	Ultra-Low Power
ULV	Ultra-Low Voltage
WI	Weak Inversion
WPE	Well Proximity Effect

Symbols

Av_{cl}	Closed-Loop Voltage Gain
Av_{cl_0}	Low Frequency Closed-Loop Voltage Gain
Av_{dm}	Differential-Mode Voltage Gain
Av_{dm_0}	Low Frequency Differential-Mode Voltage Gain
Av_{cm}	Common-Mode Voltage Gain
Av_{cm_0}	Low Frequency Common-Mode Voltage Gain
Av_{vdd}	Positive Power Supply Voltage Gain
Av_{vss}	Negative Power Supply Voltage Gain
C_{bs}	Bulk to Source Capacitance
C_{db}	Drain to Bulk Capacitance
C_{gb}	Gate to Bulk Capacitance
C_{gd}	Gate to Drain Capacitance
C_{gs}	Gate to Source Capacitance
C_i	Input Capacitance
C_{io}	Input to Output Capacitance
C_L	Load Capacitance
C_o	Output Capacitance
C_{ob}	Output to Bulk Capacitance
C_{ox}	Gate Oxide Capacitance
ΔV_T	Threshold Voltage Variation
Δg_m	Transconductance Safety Margin
e_{rr}	Percent Error
f_c	Solution Evaluation Cost Function
f_{cutoff}	Cutoff Frequency
ϕ_T	Thermal Voltage
ϕ_F	Fermi Level

γ	Body Effect Parameter
γ_n	MOSFET Thermal Noise Factor
g_m	Transconductance
g_{mb}	Bulk Transconductance
g_{mneg}	Equivalent Negative Transconductance
g_m/I_D	Transistor Efficiency Parameter
g_{ds}	Output Conductance
I_0	Technology Current
I_B	Bulk Current
I_{BS}	Bulk to Source Current
I_C	Channel Inversion Coefficient
I_D	Drain Current
$I_D/(W/L)$	Drain Current Density
I_G	Gate Leakage Current
I_{ref}	Reference Current
I_S	Source Current
k	Boltzmann Constant (approximately equal to 1.38×10^{-23} J/K)
L	Transistor Channel Length
Λ	Channel Length Modulation Parameter
M	Device Multiplicity
μ_0	Low Electric Field Channel Charge Carrier Mobility Parameter
n	Sub-threshold Slope Factor
n_0	Bulk Factor Parameter
ω_{GBW}	Angular Gain-Bandwidth Product Frequency
ω_p	Pole Angular frequency
ω_z	Zero Angular frequency
p_i	Process Model Parameters
Q_{filter}	Filter Quality Factor
s	Laplace Frequency ($s \approx j.\omega$)
$\overline{V_n^2}$	Noise Power Source
V_B	Bias Voltage
V_{BS}	Bulk to Source Voltage
V_{CM}	Common-Mode Voltage
$V_{CM_{DC}}$	DC Common-Mode Voltage
$V_{CM_{in_{min}}}$	Minimum Input Common-Mode Voltage
V_{DD}	Power Supply Voltage
$V_{DD_{min}}$	Minimum Power Supply Voltage
V_{DS}	Drain to Source Voltage
$V_{DS_{SAT}}$	Saturation Voltage
V_{GS}	Gate to Source Voltage
$V_{iCM_{DC}}$	Minimum Common-Mode Input Voltage
V_{OD}	Overdrive Voltage
$V_{O_{DC}}$	Output DC Voltage
$V_{O_{CM}}$	Output CM Voltage

V_T Threshold Voltage

V_{T0} Threshold Voltage at $V_{BS} = 0V$

V_{T_p} PMOS Transistor Threshold Voltage

T Temperature in Kelvin (K)

W Transistor Channel Width

Chapter 1
Introduction

1.1 Overview

The number of electronics devices connected to the internet is increasing exponentially over the past years. In the past, each person used to have only one or two connected devices, such as the personal computer and a smartphone. Recently, a large number of personal gadget together with the home automation platforms and the internet of Things (IoT) concepts have increased the number of connected devices by person. It is expected that each person will have around 40 connected things by 2030, resulting in 350 billion of IoT devices in operation. It is also estimated that the IoT will generate a market of $16 trillion in the same year [27]. Due to this, IoT becomes one of the main subjects of the microelectronics research centers and industry nowadays.

The application of the IoT devices is not limited to personal gadget, but can be used in several applications, from the biomedical to the precision agriculture [9]. Also is a great potential of the use of Artificial Intelligence applications embedded to the IoT devices [27]. The massive number of wirelessly connected devices and it exponential growth have increased the demand for short-range RF transceivers. In some applications, such as the body implantable devices and the remote wireless sensor networks (WSN), there is a significant trade-off between the required long lifetime and the reduced energy density availability. This is supported by the challenging, sometimes impracticable, device replacement and the low energy density provided by small batteries or the energy harvesting capability [12]. As the RF transceiver is one of the most power hungry circuit of an IoT device, consuming about 90% of the total energy [25], the key solution to address the trade-off is the development of low energy RF transceivers.

Several solutions have been proposed in the literature to make the design of low energy transceivers possible. The first of all strategies is the communication standard improvements to relax the hardware requirements in favor of the low power

L. C. Severo, W. A. M. Van Noije, *Ultra-low Voltage Low Power Active-RC Filters and Amplifiers for Low Energy RF Receivers*, https://doi.org/10.1007/978-3-030-90103-5_1

dissipation [7], as example, the Bluetooth Low Energy (BLE) standard [5]. The physical level simplification of the BLE standard made possible and practical several new RF transceiver architectures, such as digital-intensive circuits transceivers [23], reducing the number of RF active blocks [6, 28] and operating with ultra low-voltage (ULV) supply [6, 13, 23, 40, 47, 49].

Some of the new architectural implementations, as the all-digital transceivers should be implemented in advanced CMOS process ($\leq$40 nm) because of the need for faster switches and lower parasitic capacitances [23]. In contrast, the strategies based on removing some active-RF classical building blocks, such as the low noise amplifiers (LNA), and the ULV operation can also be implemented in low-cost sub-micron CMOS processes. It can be a interesting strategy to reduce the IC cost by using a More than Moore approach with the system in package (SiP) implementations [37], using different CMOS processes to implement a complete device. Additionally, the low voltage operation offers the opportunity to increase the IoT device lifetime, by using high-efficiency low conversion factor DC-DC converters, on the battery [26] and energy harvesting [1, 48] powered circuits, respectively.

The ULV operation, for voltage supplies lower than 0.5 V [8], is also very useful for digital circuits, in which an optimal supply voltage level can be adopted to reduce the dynamic power dissipation to the level of the leakage power dissipation to reach the Minimal Energy Point (MEP) operation [2, 36].

1.2 Low Energy RF Receivers

The development of low energy RF transceivers is the key solution for improving the lifetime of the IoT devices. The receiver (RX) part of a transceiver should be able of picking-up the RF signal received by the antenna, at the desired frequency band, and to process the received signal in order to give to the digital processing block the received information.

Table 1.1 shows the state-of-the-art 2.4 GHz BLE receivers from the literature. In general, the power dissipation have been reduced by the time, reaching it minimal level around 300 µW [20, 21, 43, 47] using low voltage operation and advanced CMOS processes. In these cases, the receiver main specifications are equivalent noise figure (NF) higher than 5.5 dB, input third-order intercept point (IIP3) lower than -7.5 dBm and a reasonable gain range from 30 dB to 60 dB.

The state-of-the-art low energy RXs from the literature are mainly composed by continuous-time quadrature topologies. Figure 1.1 shows the block diagram of a typical Low-IF topologies. The RF signal received by the antenna is coupled to the receiver front-end by using an input matching network (IMN) to maximize the power transference from the antenna to the RX.

The front-end first active block is the low noise amplifier (LNA) that is used to amplify the received signal with low noise insertion. The LNA circuits usually are the most power hungry circuit in an RX. It is mainly due to the low noise figure (NF)

Table 1.1 The state-of-the-art 2.4 GHz Bluetooth LE RF receivers

Reference	Process [nm]	Architecture	Voltage [V]	Power [mW]	NF [dB]	IIP3 [dBm]	Gain [dB]
T-MTT'13 [28]	130	Zero-IF[a]	1.0	1.1	16.1	2.9	–
JSSC'13 [45]	130	Sliding-IF	1	4.8	6.0	–	–
ISSCC'13 [24]	90	Sliding-IF	1.2	3.6	6.0	−19	76
ISSCC'13 [50]	65	Low-IF	0.3	1.3	6.1	−21.5	83
JETCAS'14 [6]	65	Zero-IF[a]	0.85	0.55	9.6	−3	41
JSSC'15 [40]	130	Low-IF	0.8	0.6	15.1	−15.8	56.1
JSSC'15 [32]	55	Low-IF	0.9–3.3	11.2	–	–	–
ISSCC'15 [38]	40	Sliding-IF	1.1	6.3	6.5	–	–
ISSCC'15 [25]	40	Sliding-IF	1.0	3.3	–	–	–
ESSCIRC'16 [44]	40	Sliding-IF	0.92–1.1	5.3	–	–	–
JSCC'17 [22]	28	High-IF DT	1.0	2.75	6.5	−19	46
CICC'17 [41]	40	Zero-IF	1	0.98	5.2	−19.7	72
ISSCC'18 [13]	40	Zero-IF	0.8	2.3	5.9	–	–
IMS'18 [23]	28	Low-IF DT	0.55[b]	1	7.9	−13.6	37
JSSC'18 [47]	28	Low-IF[c]	0.18→0.30[d]	0.38→1.31	8.8→11.3	−12.5→4.8	
NEWCAS'18 [35]	180	Zero-IF	0.8	0.55	14	−16	42
A-SSCC'19 [20]	22	Low-IF	0.55	0.33	9.4	−8	32.3
SSCL'19 [21]	28	Low-IF	0.9	0.35	6.5	−8	53.3
ISSCC'20 [43]	22	Zero-IF	0.7	0.37	5.5	−7.5	61

[a]LNA-less front-end
[b]Core voltage
[c]Without baseband filters and amplifiers
[d]DC-DC converter input voltage

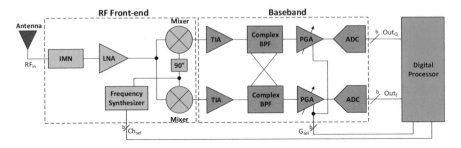

Fig. 1.1 A typical low-IF RF receiver architecture

Table 1.2 State-of-the-arts low-voltage low-energy LNAs

Reference	Process [nm]	Voltage [V]	Power [µW]	NF [dB]	Gain [dB]	IIP3 [dBm]	Frequency [GHz]
RFIC'11 [42]	130	0.4	60	5.3	13.1	−12.2	2.4
ISCAS'13 [29]	90	0.4	410	4.5 ↔ 5.3	15	−2 ↔ −7	3.2 ↔ 10
JSSC'16 [30]	130	0.4	160	4.5	13	−10	0.6 ↔ 3.1
ISCAS'17 [18]	40	0.8	30	3.3	14.2	−13.2	2.4
TCAS-I'18 [19]	40	0.18	30	5.2	14	−8.6	2.4
TCAS-II'20 [17]	180	0.6	600	4	4 ↔ 10	0	2.8

requirements of the classical communications standards in order to present a high sensitivity. However, due to the relaxed NF requirement of the BLE standard, the LNA can be designed with noise figure (NF) around 5 dB and it power consumption can be optimized by reducing it supply voltage. Table 1.2, show the state-of-the-art 2.4 GHz low-voltage LNAs used in low energy receivers. The lowest power dissipation is reached by the LNAs presented in [18] and [19] that dissipates only 30 µW when powered with 0.8 V and 0.18 V, respectively.

After the LNA, at the RX block diagram, there is the frequency down-conversion stage performed that is composed by the quadrature mixer and the frequency synthesizer. The RF signal is translated to the baseband frequency around 1–3 MHz in Low-IF architectures, or to 0 Hz in Zero-IF topologies. The output frequency is defined by the frequency synthesizer block in order to perform the channel selection. In the low-voltage implementations, passive mixer and transimpedance amplifiers (TIA) are commonly used to the down-conversion process [34] and digitally-controlled oscillator (DCO) is preferred to implement the frequency synthesizer [31, 48].

After the down-conversion process, the analog signal should be filtered and amplified. The filtering is performed with a complex band-pass filter in order to select the desired channel and to remove the image signals [10]. In a Zero-IF implementation the filtering is performed by a simple low-pass filter (LPF). The signal amplification is done by a Programmable-Gain Amplifier (PGA) that has it gain programmed by the digital processor or a automatic gain control (AGC) in order to keep a reasonable output level. At the end, a analog to digital converter is used to interface the signal to the DSP. The ADC circuit is commonly implemented using a successive approximation register (SAR) architecture to reduce the number of active blocks and, consequently, to reduce the power dissipation [16].

1.3 Baseband Filters and Amplifiers

Classically, the power optimization was mainly motivated for the high-frequency circuits, such as the LNA and the voltage controlled oscillator (VCO). As the baseband circuits used to present much lower power dissipation in comparison to the RF parts, it implementation were based on using multi-stage operational amplifiers (opamps). The classical baseband filters implementation present the power/pole relation over 500 μW per pole [11]. Thus, a simple second-order biquadratic filter can easily overpass the 1 mW of power dissipation. As shown in Sect. 1.2, the modern LNAs can dissipate as low as 30 μW is. So, there is a great motivation of power optimization also at the base-band circuits.

The continuous time active-RC topologies are preferred to implement the baseband filter and amplifiers because of its higher linearity in comparison to the g_m-C topologies at the ULV range [4, 49]. These circuits have the operational amplifier as the main active block and due to the required gain-bandwidth product (GBW), the ULV topologies tend to present more than 100 μW of power dissipation per pole, even using single-stage amplifiers [33, 46] and optimized designs [3].

The active-RC filters implementations are based on using operational amplifiers with resistors and capacitors to perform the feedback and to control frequency characteristics. Figure 1.2a and b show the basic cell employed to implement active-RC low-pass filters with single-ended or fully-differential (balanced) operational amplifiers, respectively. The closed-loop nature of the active-RC implementation reduces the swing voltage at the amplifier inputs and improve its linearity, as illustrated by the sinusoidal waves in Fig. 1.2. This characteristic makes the active-RC topologies better than the g_m-C topologies to operate at the ULV range. The use of balanced amplifiers instead of using single-ended amplifiers also is preferred to implement ULV circuits since they have twice the output voltage swing of the single-end implementations, as illustrated in Fig. 1.2. Furthermore, the use of fully-differential filter topologies facilitates the implementation of high order integrator-based active filters, considering that both the negative and positive signals are available [39]. In contrast to this, a common-mode feedback circuit (CMFB) is required to control the output common-mode signal of the balanced amplifier. The design of this circuit can increase the operational amplifier power dissipation considerably if the conventional CMFB circuits topologies are used at the ULV range [14, 15].

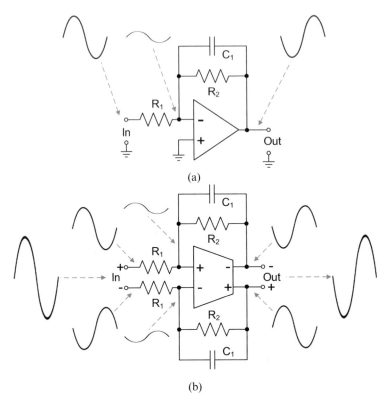

Fig. 1.2 The basic cell of an active-RC filter implementation using operational amplifiers:
(**a**) single-ended and (**b**) fully-differential

1.4 Book Organization

In this book, novel solutions are presented and analyzed for the implementation of
low-power baseband filters and amplifiers operating at the ultra-low voltage range.

In Chap. 2, the CMOS transistor low-voltage operations are analyzed employing
some I–V measurements. Based on that the voltage and power limits of the low-
voltage operational amplifiers are also analyzed. A review of the modern strategies
used to reduce the power-dissipation and to compensate for the Process-Voltage-
Temperature (PVT) variability is also presented for ULV amplifiers.

In Chap. 3, the use of single-stage amplifiers for the active-RC filter is analyzed
and some compensations techniques are presented. Also, some 0.4 V low-power
inverter-based amplifier topologies are presented.

Chapter 4, presents a design tool that makes easy and fasts the design of ULV
circuits using operation point analysis. In this chapter, some semi-automatic design
flows are proposed at the schematic level.

In Chap. 5, both the amplifier topologies presented in Chap. 3 and the methodologies presented in Chap. 4 are used to design filters and programmable amplifiers for low energy receivers. All the steps from design to measurement are detailed.

Finally, Chap. 6 concludes this book and presents some promising opportunities for this area of researches.

References

1. M. Abdelfattah, M. Swilam, B. Dupaix, S. Smith, A. Fayed, W. Khalil. An on-chip resonant-gate-drive switched-capacitor converter for near-threshold computing achieving 70.2% efficiency at 0.92 A/mm2 current density and 0.4 V output, in *2018 IEEE International Solid-State Circuits Conference (ISSCC)*, San Francisco (IEEE, New York, 2018), pp. 438–440. https://doi.org/10.1109/ISSCC.2018.8310372
2. M. Alioto, Ultra-low power VLSI circuit design demystified and explained: a tutorial. IEEE Trans. Circ. Syst. I: Regul. Pap. **59**(1), 3–29 (2012). https://doi.org/10.1109/TCSI.2011.2177004
3. H.A. Alzaher, F.S. Tasadduq, F.S. Al-Ammari, Optimal low power complex filters. IEEE Trans. Circ. Syst. I: Regul. Pap. **60**(4), 885–895 (2013). https://doi.org/10.1109/TCSI.2012.2209293
4. A. Balankutty, S.A. Yu, Y. Feng, P.R. Kinget, A 0.6-V zero-IF/low-IF receiver with integrated fractional-N synthesizer for 2.4-GHz ISM-band applications. IEEE J. Solid-State Circ. **45**(3), 538–553 (2010). https://doi.org/10.1109/JSSC.2009.2039827
5. Bluetooth-SIG. Bluetooth 5 Core Specification, 2016. https://www.bluetooth.com/specifications/%0Abluetooth-core-specification
6. C. Bryant, H. Sjoland, A 0.55 mW SAW-less receiver front-end for bluetooth low energy applications. IEEE J. Emerg. Select. Top. Circ. Syst. **4**(3), 262–272 (2014). https://doi.org/10.1109/JETCAS.2014.2337153
7. K.H. Chang, Bluetooth: a viable solution for IoT? IEEE Wirel. Commun. **21**(6), 6–7 (2014). https://doi.org/10.1109/MWC.2014.7000963
8. S. Chatterjee, Y. Tsividis, P. Kinget, 0.5-V analog circuit techniques and their application in OTA and filter design. IEEE J. Solid-State Circ. **40**(12), 2373–2387 (2005). https://doi.org/10.1109/JSSC.2005.856280
9. L. Columbus, 2017 Roundup Of Internet Of Things Forecasts, 2017. https://www.forbes.com/sites/louiscolumbus/2017/12/10/2017-roundup-of-internet-of-things-forecasts/#3c5378d11480
10. L. Compassi-Severo, W. Van Noije, A 0.4-V 10.9-μW/pole third-order complex BPF for low energy RF receivers. IEEE Trans. Circ. Syst. I: Regul. Pap. **66**(6), 2017–2026 (2019). ISSN:15580806. https://doi.org/10.1109/TCSI.2019.2906206
11. M. De Matteis, S. D'Amico, A. Baschirotto, A 0.55 V 60 dB-DR fourth-order analog baseband filter. IEEE J. Solid-State Circ. **44**(9), 2525–2534 (2009) https://doi.org/10.1109/JSSC.2009.2024801
12. S.M. Demir, F. Al-Turjman, A. Muhtaroglu, Energy scavenging methods for WBAN applications: a review. IEEE Sens. J. **18**(16), 6477–6488 (2018). https://doi.org/10.1109/JSEN.2018.2851187
13. M. Ding, X. Wang, P. Zhang, Y. He, S. Traferro, K. Shibata, M. Song, H. Korpela, K. Ueda, Y.-h. Liu, C. Bachmann, K. Philips, A 0.8 V 0.8 mm2 bluetooth 5/BLE digital-intensive transceiver with a 2.3 mW phase-tracking RX utilizing a hybrid loop filter for interference resilience in 40 nm CMOS, in *2018 IEEE International Solid - State Circuits Conference - (ISSCC)*, ed. by IEEE, San Francisco, 2018 (IEEE, New York, 2018), pp. 446–448. https://doi.org/10.1109/ISSCC.2018.8310376

14. J.F. Duque-Carrillo, Control of the common-mode component in CMOS continuous-time fully differential signal processing. Analog Integr. Circ. Signal Process. **4**(2), 131–140 (1993) https://doi.org/10.1007/BF01254864

15. R. Harjani, R.K. Palani, Design of PVT tolerant inverter based circuits for low supply voltages. Proc. Custom Integr. Circ. Conf. **1**(3), 2015. https://doi.org/10.1109/CICC.2015.7338424

16. H.D. Hernandez, L.C. Severo, W.A.M.V. Noije, 0.5 V 10 MS/s 9-bits asynchronous SAR ADC for BLE receivers in 180 nm CMOS technology, in *Proceedings of the IEEE Systems on Chip Conference (SOCC2018)*, Washington DC, 2018 (IEEE, New york, 2018)

17. J.-Y. Hsieh, K.-Y. Lin, A 0.6-V low-power variable-gain LNA in 0.18-$\umu$m CMOS technology. IEEE Trans. Circ. Syst. II: Express Briefs **67**(1), 23–26 (2020). ISSN:1549-7747. https://doi.org/10.1109/TCSII.2019.2902301. https://ieeexplore.ieee.org/document/8654671/

18. E. Kargaran, D. Manstretta, R. Castello, A 30 μW, 3.3 dB NF CMOS LNA for wearable WSN applications. *Proceedings - IEEE International Symposium on Circuits and Systems*, 2017, pp. 1–4. ISSN:02714310. https://doi.org/10.1109/ISCAS.2017.8050597

19. E. Kargaran, D. Manstretta, R. Castello, A sub-1 V, 220 μw receiver frontend for wearable wireless sensor network applications, in *Proceedings - IEEE International Symposium on Circuits and Systems*, 2018 May (2018), pp. 1–5. ISSN:02714310. https://doi.org/10.1109/ISCAS.2018.8351643

20. E. Kargaran, C. Bryant, D. Manstretta, J. Strange, R. Castello, A sub-0.6 V, 330 μw, 0.15 mmˆ2 receiver front-end for bluetooth low energy (BLE) in 22 nm FD-SOI with zero external components, in *Proceedings - 2019 IEEE Asian Solid-State Circuits Conference, A-SSCC 2019*, vol. 6 (IEEE, New York, 2019), pp. 169–172. ISBN:9781728151069. https://doi.org/10.1109/A-SSCC47793.2019.9056899

21. E. Kargaran, B. Guo, D. Manstretta, R. Castello, A sub-1-V, 350-$\umu$ W, 6.5-dB integrated NF low-IF receiver front-end for IoT in 28-nm CMOS. IEEE Solid-State Circ. Lett. **2**(4), 29–32 (2019). ISSN:2573-9603. https://doi.org/10.1109/LSSC.2019.2917870. https://ieeexplore.ieee.org/document/8718256/

22. F.-W. Kuo, S. Binsfeld Ferreira, H.-N.R. Chen, L.-C. Cho, C.-P. Jou, F.-L. Hsueh, I. Madadi, M. Tohidian, M. Shahmohammadi, M. Babaie, R.B. Staszewski, A bluetooth low-energy transceiver with 3.7-mW all-digital transmitter, 2.75-mW high-IF discrete-time receiver, and TX/RX switchable on-chip matching network. IEEE J. Solid-State Circ. **52**(4), 1144–1162 (2017). https://doi.org/10.1109/JSSC.2017.2654322

23. F.-w. Kuo, S.B. Ferreira, R. Chen, L.-C. Cho, C.-P. Jou, M. Chen, M. Babaie, R.B. Staszewski, Towards ultra-low-voltage and ultra-low-power discrete-time receivers for Internet-of-Things, in *2018 IEEE/MTT-S International Microwave Symposium - IMS*, 2018, pp. 1211–1214

24. Y.H. Liu, X. Huang, M. Vidojkovic, A. Ba, P. Harpe, G. Dolmans, H. De Groot, A 1.9 nJ/b 2.4 GHz multistandard (Bluetooth low energy/Zigbee/IEEE802.15.6) transceiver for personal/body-area networks. *Digest of Technical Papers - IEEE International Solid-State Circuits Conference*, vol. 56 (2013), pp. 446–447. https://doi.org/10.1109/ISSCC.2013.6487808

25. Y.H. Liu, C. Bachmann, X. Wang, Y. Zhang, A. Ba, B. Busze, M. Ding, P. Harpe, G.-J. van Schaik, G. Selimis, H. Giesen, J. Gloudemans, A. Sbai, L. Huang, H. Kato, G. Dolmans, K. Philips, H. de Groot, A 3.7 mW-RX 4.4 mW-TX fully integrated Bluetooth Low-Energy/IEEE802.15.4/proprietary SoC with an ADPLL-based fast frequency offset compensation in 40 nm CMOS, in *2015 IEEE International Solid-State Circuits Conference - (ISSCC) Digest of Technical Papers*, San Francisco, 2015 (IEEE, New York, 2015), pp. 1–3. https://doi.org/10.1109/ISSCC.2015.7063013

26. Y.-H. Liu, V.K. Purushothaman, C. Lu, J. Dijkhuis, R.B. Staszewski, C. Bachmann, K. Philips, A 770 pJ/b 0.85 V 0.3 mm 2 DCO-based phase-tracking RX featuring direct demodulation and data-aided carrier tracking for IoT applications, in *2017 IEEE International Solid-State Circuits Conference (ISSCC)*, vol. 60, San Francisco, 2017 (IEEE, New york, 2017), pp. 408–409. https://doi.org/10.1109/ISSCC.2017.7870434

27. K.-H.L. Loh, 1.2 fertilizing AIoT from roots to leaves, in *2020 IEEE International Solid- State Circuits Conference - (ISSCC)* (IEEE, New York,2020), pp. 15–21. ISBN:978-1-7281-3205-1. https://doi.org/10.1109/ISSCC19947.2020.9062950. https://ieeexplore.ieee.org/document/9062950/

28. J. Masuch, M. Delgado-Restituto, A 1.1-mW-RX -81.4-dBm sensitivity CMOS transceiver for Bluetooth lOW energy. IEEE Trans. Microw. Theory Tech. **61**(4), 1660–1673 (2013). https://doi.org/10.1109/TMTT.2013.2247621

29. M. Parvizi, K. Allidina, F. Nabki, M. El-Gamal, A 0.4 V ultra low-power UWB CMOS LNA employing noise cancellation. *Proceedings - IEEE International Symposium on Circuits and Systems* (2), 2369–2372 (2013). ISSN:02714310. https://doi.org/10.1109/ISCAS.2013.6572354

30. M. Parvizi, K. Allidina, M.N. El-Gamal, Short channel output conductance enhancement through forward body biasing to realize a 0.5 V 250 μW 0.6–4.2 GHz current-reuse CMOS LNA. IEEE J. Solid-State Circ. **51**(3), 574–586 (2016) ISSN:00189200. https://doi.org/10.1109/JSSC.2015.2504413

31. N. Pourmousavian, F.W. Kuo, T. Siriburanon, M. Babaie, R.B. Staszewski, A 0.5-V 1.6-mW 2.4-GHz fractional-N all-digital PLL for Bluetooth le with PVT-insensitive TDC using switched-capacitor doubler in 28-nm CMOS. IEEE J. Solid-State Circ. **53**(9), 2572–2583 (2018). https://doi.org/10.1109/JSSC.2018.2843337

32. J. Prummel, M. Papamichail, J. Willms, R. Todi, W. Aartsen, W. Kruiskamp, J. Haanstra, E. Opbroek, S. Rievers, P. Seesink, J. Van Gorsel, H. Woering, C. Smit, A 10 mW Bluetooth low-energy transceiver with On-chip matching. IEEE J. Solid-State Circ. **50**(12), 3077–3088 (2015). https://doi.org/10.1109/JSSC.2015.2469674

33. A. Rasekh, M. Sharif Bakhtiar, Design of low-power low-area tunable active RC filters. IEEE Trans. Circ. Syst. II: Express Briefs **65**, 6–10 (2017). https://doi.org/10.1109/TCSII.2017.2658635

34. B. Razavi, *RF Microelectronics*, vol. 53, 2th edn. (Prentice Hall, New York, 2012). https://doi.org/10.1017/CBO9781107415324.004

35. S.S. Regulagadda, S. Nagaveni, A. Dutta, A 550-MW, 2.4-GHz ZigBee/BLE receiver front end for IoT applications in 180-nm CMOS, in *2018 16th IEEE International New Circuits and Systems Conference, NEWCAS 2018*, 2018, pp. 48–52. https://doi.org/10.1109/NEWCAS.2018.8585629

36. N. Reynders, W. Dehaene, *Ultra-Low-Voltage Design of Energy-Efficient Digital Circuits*, 1st edn. (Springer, New York, 2015)

37. V. Roche, Semiconductor innovation: is the party over, or just getting started?, in *2018 IEEE International Solid - State Circuits Conference - (ISSCC)* (IEEE, New York, 2018), pp. 8–11. ISBN:978-1-5090-4940-0. https://doi.org/10.1109/ISSCC.2018.8310164. http://ieeexplore.ieee.org/document/8310164/

38. T. Sano, M. Mizokami, H. Matsui, K. Ueda, K. Shibata, K. Toyota, T. Saitou, H. Sato, K. Yahagi, Y. Hayashi, A 6.3 mW BLE transceiver embedded RX image-rejection filter and TX harmonic-suppression filter reusing on-chip matching network, in *Digest of Technical Papers - IEEE International Solid-State Circuits Conference*, vol. 58, 240–241 (2015). https://doi.org/10.1109/ISSCC.2015.7063015

39. A.S. Sedra, K.C. Smith, *Microeletrônica* (Pearson, São Paulo, 2007)

40. A. Selvakumar, A. Liscidini, Current-recycling complex filter for bluetooth-low-energy applications. IEEE Trans. Circ. Syst. II: Express Briefs **62**(4), 332–336 (2015). https://doi.org/10.1109/TCSII.2014.2387611

41. A.H.M. Shirazi, H.M. Lavasani, M. Sharifzadeh, Y. Rajavi, S. Mirabbasi, M. Taghivand, A 980 μW 5.2 dB-NF current-reused direct-conversion bluetooth-low-energy receiver in 40 nm CMOS, in *2017 IEEE Custom Integrated Circuits Conference (CICC)* (IEEE, New York, 2017), pp. 1–4. ISBN:978-1-5090-5191-5. https://doi.org/10.1109/CICC.2017.7993647. http://ieeexplore.ieee.org/document/7993647/

42. T. Taris, J.B. Begueret, Y. Deval, A 60 μW LNA for 2.4 GHz wireless sensors network applications, in *Digest of Papers - IEEE Radio Frequency Integrated Circuits Symposium* (V), (2011), pp. 1–4. ISSN:15292517. https://doi.org/10.1109/RFIC.2011.5940633

43. B.J. Thijssen, E.A.M. Klumperink, P. Quinlan, B. Nauta, 30.4 A 370 μW 5.5 dB-NF BLE/BT5.0/IEEE 802.15.4-compliant receiver with >63 dB adjacent channel rejection at >2 channels offset in 22 nm FDSOI, in *2020 IEEE International Solid- State Circuits Conference - (ISSCC)* (IEEE, New York, 2020), pp. 466–468. ISBN:978-1-7281-3205-1. https://doi.org/10.1109/ISSCC19947.2020.9062973. https://ieeexplore.ieee.org/document/9062973/

44. X. Wang, J. Van den Heuvel, G.-J. van Schaik, C. Lu, Y. He, A. Ba, B. Busze, M. Ding, Y.-H. Liu, N. Winkel, M. Wildeboer, C. Bachmann, K. Philips, A 0.9–1.2 V supplied, 2.4 GHz Bluetooth low energy 4.0/4.2 and 802.15.4 transceiver SoC optimized for battery life, in *ESSCIRC Conference 2016: 42nd European Solid-State Circuits Conference*, vol. 2016, October, Lausanne, 2016 (IEEE, New York, 2016), pp. 125–128. https://doi.org/10.1109/ESSCIRC.2016.7598258

45. A.C.W. Wong, M. Dawkins, G. Devita, N. Kasparidis, A. Katsiamis, O. King, F. Lauria, J. Schiff, A.J. Burdett, A 1 V 5 mA multimode IEEE 802.15.6/bluetooth low-energy WBAN transceiver for biotelemetry applications. IEEE J. Solid-State Circ. **48**(1), 186–198 (2013). https://doi.org/10.1109/JSSC.2012.2221215

46. L. Ye, C. Shi, H. Liao, R. Huang, Y. Wang, Highly power-efficient active-RC filters with wide bandwidth-range using low-gain push-pull Opamps. IEEE Trans. Circ. Syst. I: Regul. Pap. **60**(1), 95–107 (2013)

47. H. Yi, W.-H. Yu, P.-I. Mak, J. Yin, R.P. Martins, A 0.18-V 382-μW Bluetooth low-energy receiver front-end with 1.33-nW sleep power for energy-harvesting applications in 28-nm CMOS. IEEE J. Solid-State Circ. **53**(6), 1618–1627 (2018)

48. J. Yin, S. Yang, H. Yi, W.-H. Yu, P.-I. Mak, R.P. Martins, A 0.2 V energy-harvesting BLE transmitter with a micropower manager achieving 25% system efficiency at 0 dBm output and 5.2 nW sleep power in 28 nm CMOS, in *2018 IEEE International Solid - State Circuits Conference - (ISSCC)*, pp. 450–452 (IEEE, New York, 2018). https://doi.org/10.1109/ISSCC.2018.8310378.

49. F. Zhang, Y. Miyahara, B.P. Otis, Design of a 300-mV 2.4-GHz receiver using transformer-coupled techniques. IEEE J. Solid-State Circ. **48**(12), 3190–3205 (2013) https://doi.org/10.1109/JSSC.2013.2280835

50. F. Zhang, K. Wang, J. Koo, Y. Miyahara, B. Otis, A 1.6 mW 300 mV-Supply 2.4 GHz Receiver with −94 dBm sensitivity for energy-harvesting applications, in *2013 IEEE International Solid-State Circuits Conference* (2013), pp. 456–458. ISBN:9781467345163

Chapter 2
ULV and ULP Operational Amplifiers for Active-RC Filters

2.1 Low Voltage Operation of CMOS Transistors

The electrical characteristics of the CMOS transistors are dependent on the channel Width (W) and Length (L), the fabrication process and on the bias voltage. The W and L parameters are the designer free variables, while the bias voltage is limited by the used power supply voltage level and by the circuit topology.

In the following subsections, the main characteristics of the CMOS transistor operation at low voltage are analyzed using some experimental device I–V curves from a 130 nm CMOS process. The I-V curves are obtained with the microprobe measurements of long channel transistors M1 (Low-V_T) and M2 (standard-V_T), and on variable channel length transistors M3a, M3b and M3c (Low-V_T), shown in Fig. 2.1. Although the transistor bulk terminal is the best terminal voltage to be adopted as a reference for the transistor characteristics analysis [26], this analysis uses a common-source reference to make easier the analysis of the bulk forward bias.

2.1.1 Current Density and Channel Inversion Level

The CMOS transistor drain current (I_D) is directly dependent on the gate to source (V_{GS}) and on the drain to source (V_{DS}) voltages. The I_D is also directly proportional to the channel width (W) and inversely proportional to the channel length (L). Thus, the transistor drain current density can be evaluated as the $I_D/(W/L)$ ratio and it is one of the most important parameters for the low-voltage operation. The lower $I_D/(W/L)$ is, the higher should be the transistor W/L aspect ratio do present the target drain current. Figure 2.2 shows the measured drain current density ($I_D/(W/L)$) in linear and logarithms scales for the two CMOS transistor with

© The Author(s), under exclusive license to Springer Nature Switzerland AG 2022   11
L. C. Severo, W. A. M. Van Noije, *Ultra-low Voltage Low Power Active-RC Filters and Amplifiers for Low Energy RF Receivers*,
https://doi.org/10.1007/978-3-030-90103-5_2

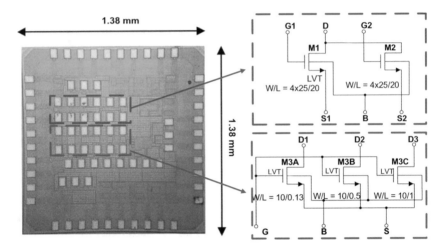

Fig. 2.1 Test Transistor used for experimental analysis of the CMOS transistor operating at low voltage

aspect to ratio of 100/20 and two different threshold voltages (V_T). The V_{GS} voltage controls the channel inversion level while the V_{DS} defines the triode and saturation regions. The border between the triode and saturation regions is defined by the saturation voltage ($V_{DS_{SAT}}$) that is analyzed in Sect. 2.1.2.

For long channel devices, at the saturation region, the drain current density is most changed by the channel inversion level. The channel inversion can be classified as weak (WI), moderated (MI) and strong (SI) inversion levels, according to the current conduction mechanism. At the WI the current conduction is dominated by charge carriers diffusion in the channel region and $I_D/(W/L)$ can be estimated by using Eq. (2.1). On the other hand, at the SI the current conduction mechanism is dominated by charge carriers drift, and the current density can be estimated by using the classical quadratic model of Eq. (2.3). In the MI level, both the drift and the diffusion current conduction mechanism are present, and it can be modeled using charge-based continuous equations, as presented in [26] and [29]. The technology current (I_0), used in Eq. (2.1), can be calculated using Eq. (2.2).

$$I_D/(W/L) = I_0.\exp\left(\frac{V_{GS}-V_T}{n.\phi_T}\right).\left[1-\exp\left(-\frac{V_{DS}}{\phi_T}\right)\right] \approx I_0.\exp\left(\frac{V_{GS}-V_T}{n.\phi_T}\right)$$
(2.1)

$$I_0 = 2.n_0.\mu_0.C_{ox}.\phi_T^2$$
(2.2)

$$I_D/(W/L) = \frac{\mu_0.C_{ox}}{2}.(V_{GS}-V_T)^2.(1+\lambda.V_{DS}) \approx \frac{\mu_0.C_{ox}}{2}.(V_{GS}-V_T)^2$$
(2.3)

where: V_T is the device threshold voltage parameter, n is the sub-threshold slope factor, ϕ_T is the thermal voltage, n_0 is the bulk factor, μ_0 is the low electric field

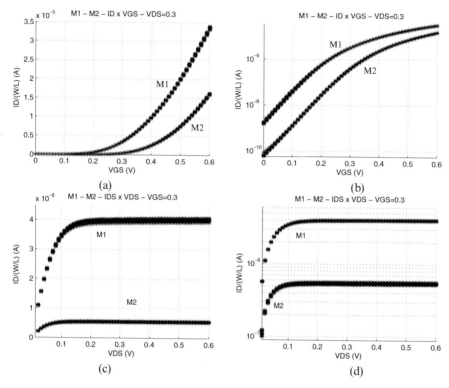

Fig. 2.2   The measured drain current density $(I_D/(W/L))$ related to the V_{GS} and V_{DS} voltages of the test transistors M1 and M2: (**a**) $I_D/(W/L) \times V_{GS}$ linear, (**b**) $I_D/(W/L) \times V_{GS}$ logarithmic, (**c**) $I_D/(W/L) \times V_{DS}$ linear and (**d**) $I_D/(W/L) \times V_{DS}$ logarithmic

mobility parameter, C_{ox} is the oxide capacitance and λ is channel length modulation parameter.

Based on Eqs. (2.1) and (2.3) we can verify that the drain current density is directly proportional to the gate effective $(V_{G_{eff}})$ or overdrive (V_{OD}) voltage, defined as $V_{GS} - V_T$. As shown in Fig. 2.2a and b, the test transistor M1 has higher values of $I_D/(W/L)$ than M2 at the same V_{GS} voltage. It occurs due to the difference on the V_{OD} voltage since M1 is a low-V_T device with V_T equal to 250 mV and M2 is a standard-V_T device with 365 mV of V_T. The classification of WI, MI and SI are empirical, and it is given according to the V_{OD} voltage, as shown in Table 2.1. Besides V_{OD}, the inversion level coefficient $(I_c = I_D/I_0)$ and the transistor efficiency given by the g_m/I_D ratio, where g_m is the gate transconductance, can also be used to define the channel inversion level, as shown in Table 2.1 [8].

At low voltage levels, the circuit designer should deal with the reduced V_{GS} voltage that makes the transistor to operate at the weak and moderated inversion levels. At this level, due to the low current density, higher transistor aspect

Table 2.1 Empirical channel inversion level classification for a NMOS transistor

	Overdrive voltage	Inversion coefficient	Transistor efficiency
Inversion level	$V_{GS} - V_T$	$I_C = I_D/I_0$	g_m/I_D
Weak	< -60 mV	< 0.1	> 20 V^{-1}
Moderated	-60 to 200 mV	0.1 to 10	10 to 20 V^{-1}
Strong	> 200 mV	> 10	< 10 V^{-1}

ratio is needed to increase the transistor drain current. In analog circuits, high transistor aspect to ratios (W/L) also increase the device parasitic capacitances and, consequently, reduces the circuit maximum operation frequency. Thus, to use transistors with reduced V_T is frequently necessary in order to increase the channel inversion level. The V_T can be reduced by replacing the standard-V_T transistor by a Low-V_T or Zero-V_T native device [12], or by using some design techniques, such as the bulk forward bias—shown in Sect. 2.1.3—and as increasing the transistor channel length—shown in Sect. 2.1.5.

2.1.2 Saturation Voltage

Some of the low voltage circuits, such as the voltage amplifiers used in this work, should operate at the saturation region in order to achieve higher voltage gain values. As a consequence, the transistors should be biased using a V_{DS} voltage higher than the minimum voltage needed to operate in saturation. The minimal V_{DS} voltage is defined as the saturation voltage ($V_{DS_{SAT}}$), and it is represented by the border between the triode and saturation regions. Figure 2.2c and d shows the characteristics curves of the drain current density ($I_D/(W/L)$) related to the V_{DS} voltage. The transistor work at the triode region for lower values of V_{DS} and at the saturation for higher values of V_{DS}. The $V_{DS_{SAT}}$ voltage is obtained experimentally at the transition between the triode and saturation regions, after the corners shown in Fig. 2.2c and d.

The saturation voltage is also dependent on the channel inversion level and can be extracted from the $I_D \times V_{DS}$ curves. Figure 2.3a and b show the measured $V_{DS_{sat}}$ of transistors M1 and M2, respectively, related to the overdrive voltage when V_{GS} is changed from 0 to 0.6 V. The minimal values for the saturation voltage are found at the WI operation. Theoretically, the minimum $V_{DS_{SAT}}$ is defined to be equal to $4.\phi_T$ that is approximately equal to 100 mV at 300 K [19]. However, according to our measurements, it can be lower than this value, at the limit of 50 mV in the deep WI. The $V_{DS_{SAT}}$ of 100 mV is found at the regions between the WI and MI with -100 mV of V_{OD}. For higher inversion levels, $V_{DS_{sat}}$ is increased proportionally to the overdrive voltage, as classically defined.

The $V_{DS_{SAT}}$ voltage is used in Sect. 2.2.1 to analyze the minimum V_{DD} voltage of the amplifiers.

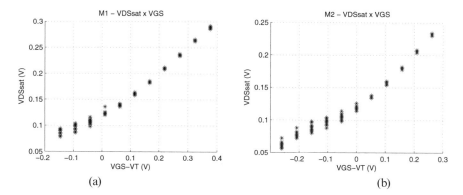

Fig. 2.3   The test transistors measured saturation voltage ($V_{DS_{sat}}$) as a function of the overdrive voltage: (**a**) M1 Low-V_T and (**b**) M2 standard-V_T

2.1.3   Bulk Forward Bias

The body effect present in bulk CMOS processes can be applied in favor to decrease the V_T of the transistor operation at low voltage circuits. The body effect on the V_T voltage can be approximately modeled using the following equation [7]:

$$V_T = V_{T0} + \gamma . \left(\sqrt{2.\Phi_F - V_{BS}} - \sqrt{2.\Phi_F} \right) \tag{2.4}$$

where: V_{T0} is the threshold voltage for $V_{BS} = 0$ V, γ is the process body effect parameter and Φ_F is the Fermi level voltage.

By forward biasing the bulk to source junction, using a positive bulk to source (V_{BS}) voltage in the NMOS transistors and a negative voltage in the PMOS transistors, the V_T can be reduced in comparison to the V_{T0} value. Figure 2.4a and b shows the measured threshold voltage according to the bulk forward bias (V_{BS}) voltage of the transistors M1 and M2. The V_T of M1 is reduced from 250 mV to about 111 mV, and the M2 V_T is reduced from 365 mV to about 221 mV when the V_{BS} voltage is changed from 0 to 0.6 V in both transistors. It is equivalent to the V_T reduction of 55.6% and 39.5% in comparison to V_{T0} for the transistor M1 and M2, respectively.

In order to analyze the increase of the channel inversion level, the drain current density versus the V_{BS} voltage with $V_{GS} = V_{DS} = 0.3$ V was measured, as shown in Fig. 2.5a. The M1 current density is moved from 3.98 μA to 14.5 μA, an increase of about 3.6 times. The transistor M2 has it current density increased about 10.4 times going from 0.55 μA to 5.71 μA. Figure 2.5b and c shows the transistor efficiency ratio g_m/I_D in relation to the V_{GS} voltage with V_{BS} equal to 0 and 0.6 V, for the M1 and M2 transistors, respectively. By using the Table 2.1 as a reference, it is possible to realize that the channel inversion level can be moved from WI to MI or from MI to SI, by changing the V_{BS} voltage from 0 to 0.6 V.

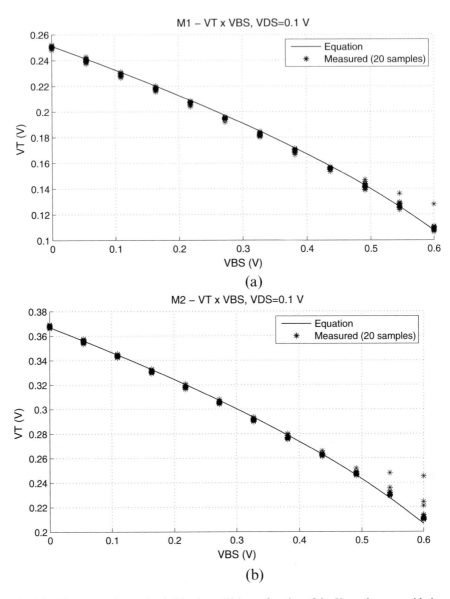

Fig. 2.4 The test transistors threshold voltage (V_T) as a function of the V_{BS} voltage, considering the measured results and Eq. (2.4) with γ equal to 0.35 $V^{1/2}$: (**a**) M1 Low-V_T and (**b**) M2 standard-V_T

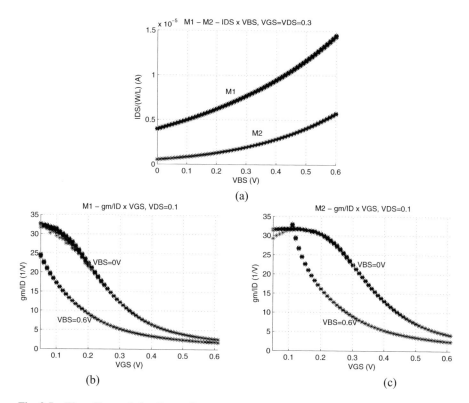

Fig. 2.5 The effects of the V_{BS} voltage on the drain current density and on the transistor efficiency: (**a**) $(I_D/(W/L))$ x V_{BS} of the M1 and M2 transistor, (**b**) g_m/I_D x V_{GS} of M1 and (**c**) g_m/I_D x V_{GS} of M2

The bulk forward bias can be extensively used in low-voltage circuits design to improve the channel inversion level without changing the V_{GS} voltage. It can also be applied to compensate for the drain current process variability by using some automatic bulk bias control. Both strategies are employed on the circuits shown in Chap. 3 of this book.

Besides the good improvements of the bulk forward bias in the inversion level, some issues should be observed. First, as the bulk to drain and bulk to source diffusions are forward biased, there is current leakage flowing from/to the bulk terminal in NMOS/PMOS transistors. Figure 2.6 shows the measured bulk current normalized to the diffusion area—W.DL—related to the V_{BS} voltage. For V_{BS} lower than 0.3 V, the leakage current density is lower than 1 $pA/\mu m^2$, and it increases exponentially from 0.3 to 0.6 V. At the maximum V_{BS} voltage equal to 0.6 V the current leakage density becomes equal to 40 nA/μm^2. The second issue is the latch-up risk due to the parasitic bipolar transistors present on the CMOS substrate. However, according to [7], these transistors are in conduction only if the bulk voltage is higher than 0.7 V, making safe to operate with V_{DD} lower than 0.6 V.

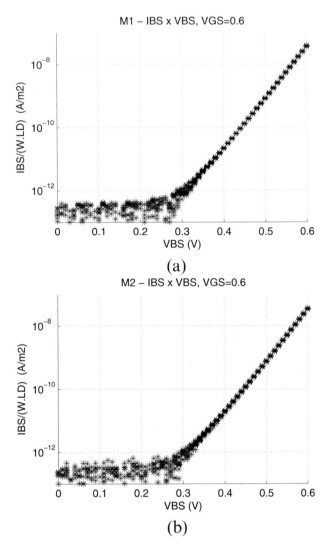

Fig. 2.6 The bulk to source leakage current, normalized to the diffusion area (W.DL), due to the bulk forward voltage: (**a**) M1 Low-V_T test transistor and (**b**) M2 standard-V_T test transistor

The circuits presented in the next chapters of this book use a maximum V_{BS} voltage of 0.4 V in order to make the current leakage density at the pA/μm^2 range, being much lower than the drain current, and to reduce the probability of latch-up.

Additionally, the use of the bulk terminal to provide a bulk forward bias, we need to use insulated bulk devices. The PMOS devices are always isolated on the P-type substrate, but the insulated bulk NMOS transistors are available only in triple-well or buried-N-well processes. Such kind of transistors are commonly available in the sub-micron and nanometer technologies at the cost of some extra masks on the IC fabrication process, and have a larger area in comparison to the standard-V_T transistors. Generally, only for the standard-VT NMOS devices are offered an insulated bulk option. Thus, the low-voltage circuits can be designed using common-bulk Low-V_T NMOS or bulk forward biased standard-V_T devices. As shown in Fig. 2.4a and b, the standard-V_T threshold voltage can be reduced to the same level of the Low-V_T device, with the V_{BS} voltage is in the range of 0.4 to 0.5 V.

2.1.4 Small-Signal Transconductances and Conductances

The common-source CMOS transistor can be analyzed by using the AC small-signal gate transconductance ($g_m = \partial I_D / \partial v_{gs}$), bulk transconductance ($g_{mb} = \partial I_D / \partial v_{bs}$) and the drain conductance ($g_{ds} = \partial I_D / \partial v_{ds}$). Figure 2.7 show the measured g_m, g_{ds} and g_{mb} curves for the test transistors M1 and M2. As the transconductances and the drain conductance are scaled with the W/L aspect ratio, the measured data are normalized to the W/L value.

Based on Eqs. (2.1) and (2.3), the $g_m/(W/L)$ at the WI and SI levels can be estimated using Eqs. (2.5) and (2.6). The g_m is directly related to the drain current at the WI and is linearly dependent on the overdrive voltage at the SI. Because of that, transistor M1 (Low-V_T) present a higher value of g_m in comparison to M2 (Standard-V_T), as shown in Fig. 2.7a. Using Eq. (2.5) we can verify that at the WI level the transistor efficiency g_m/I_D has the maximum theoretical value of $1/n.\phi_T$, around 30 V^{-1} at the room temperature, that match with the maximum values of g_m/I_D shown in Fig. 2.5b and c.

$$g_m/(W/L) = \frac{I_0}{n.\phi_T} \cdot \exp\left(\frac{V_{GS} - V_T}{n.\phi_T}\right) \approx \frac{I_D/(W/L)}{n.\phi_T} \tag{2.5}$$

$$g_m/(W/L) = \mu_0.C_{ox}.(V_{GS} - V_T) \tag{2.6}$$

The measured drain conductance (g_{ds}) is shown in Fig. 2.7b. Considering long channel devices, g_{ds} is higher at the linear region and tend to be constant at the saturation region. It is also proportional to the channel inversion level, presenting higher values in transistor M1 and lower in transistor M2. As the amplifier voltage gain is proportional to the g_m/g_{ds} ratio, higher values of voltage gain are obtained by using only saturated devices.

The bulk transconductance g_{mb} is also proportional to the inversion level and the V_{BS} value. Figure 2.7c shows the measured g_{mb} curves related to the V_{BS} voltage of transistors M1 and M2. The variation of g_{mb} is reduced in comparison to g_m, because the influence of V_{BS} on the I_D current is due to the bulk effect when the V_T is proportional to the square root variation on V_{BS}, as shown in Eq. (2.4).

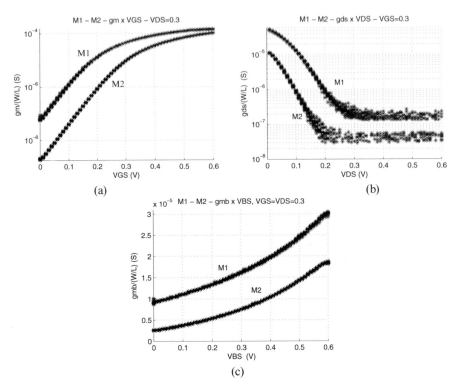

Fig. 2.7 The measured small-signal transconductance and conductance normalized to the transistor W/L aspect ration of the test transistors M1 and M2: (**a**) gate transconductance $g_m/(W/L) \times V_{GS}$, (**b**) drain conductance $g_{ds}/(W/L) \times V_{DS}$ and (**c**) bulk transconductance $g_{mb}/(W/L) \times V_{GB}$

2.1.5   Short Channel Effects

The analysis performed in the last subsection have considered two long channel devices. By reducing the channel length, several effects become important and, in general, they are worser for low-voltage operation. Figure 2.8a shows the measured $I_D/(W/L)$ related to the V_{DS} voltage for the test transistors M3A, M3B and M3C. These devices have the channel length equal to 130 nm, 500 nm and 1 μm, respectively. The saturation region slope is higher for shorter devices due to the channel length modulation effects. Due to that, the shorter devices have higher output conductances and, consequently, the g_m/g_{ds} ratio is lower. We can conclude with this figure that the current variability is higher for shorter devices and is reduced for longer devices.

The channel current density level is smaller for shorter devices as a result of the channel inversion level reduction. This effect is caused by the Halo implantation added at the transistor channel extremities. It is applied to reduce the effect of the drain-induced barrier lowering (DIBL) in the sub-micron technologies and also to

Fig. 2.8 The short channel effect on the drain current density (**a**), and the reverse short-channel effect (RSCE) on the threshold voltage (**b**)

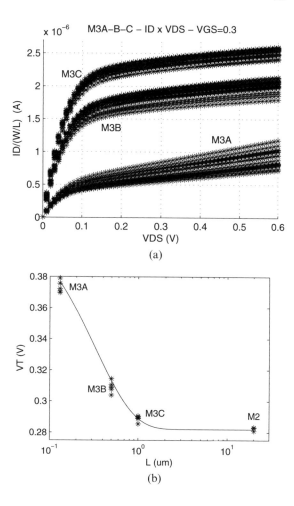

reduce the punch-through currents [29]. Due to that, shorter devices have higher V_T voltages in comparison to longer channel devices. The increase of V_T is also known as the reverse short-channel effect (RSCE) since the short channel effects without Halo implants tend to reduce the V_T. Figure 2.8b shows the measured V_T for all the low-V_T test transistors used in this work. The 130 nm channel length device has the maximum V_T of 375 mV and it is reduced to 310 mV, 290 mV and 250 mV for the 500 nm (M3B), 1 μm (M3C) and 20 μm (M1) devices, respectively. The RSCE has a higher influence at low V_{DS} voltages such as the ULV range, and it is even more significant in nanometer technologies [18].

Additionally to the V_T effect, the short-channel devices present higher mismatch variability and noise contribution when compared to longer channel transistors. Thus, the proper channel length determination in the design of low-voltage circuits is very important. It increases the design effort, since some analysis should be

implemented to allow the determination of the best channel length for each one
of the transistors present on a circuit.

2.2 ULV Operational Amplifiers

The operational amplifier is the main active building block of the active-filters. Thus,
all the filter specifications, including the power dissipation, are given as a function
of the amplifier characteristics.

In this section, the strategies to design ULV and ULP operational amplifiers are
presented and previous reported amplifiers and design solutions from the literature
are analyzed.

2.2.1 The Minimum Operation Voltage for Amplifiers

The minimum power supply voltage of a voltage amplifier is dependent on the
circuit topology and on the voltage swing desired for the circuit application. In
general, the highest voltage gain specification can be reached when all the transistors
are operating at the saturation region due to the reduced values of the output
conductance (g_{ds}). The saturation voltage ($V_{DS_{SAT}}$) of a transistor is dependent on
the channel inversion level, and it does not follow the technology scaling, as shown
in Sect. 2.1.2. Due to this limitation, the amplifiers have higher operation voltage
limits than other circuits, such as oscillators and digital inverters that can operate
with supply voltages below 100 mV [3].

Figure 2.9a shows a conventional CMOS three-stacked transistors fully-
differential amplifier. It is composed of a differential pair (M_{1A} and M_{1B}), a current
source active load (M_{2A} and M_{2B}) and a common-mode current source bias (M_{3A},
M_{3B} and I_{ref}). Transistors $M_{2A/B}$ and $M_{3A/B}$ have their gate terminals voltage
controlled by the bias voltage V_b and the bias current I_{ref} in order to present the
target drain current even with the presence of process and temperature variations.
On the other hand, the differential pair transistors are not compensated by the gate
voltage since the inputs of the amplifier have a constant DC input common-mode
voltage ($V_{iCM_{DC}}$). Thus, to make the overdrive voltage ($V_{OD} = V_{GS} - V_T$) of
$M_{1A/B}$ approximately constant, the source terminal voltage should be adjusted
through the V_{DS} voltage of transistor M_{3B}. Equation (2.7) can be written to give
the minimum supply voltage of this amplifier, considering that all the transistors are
in saturation and have the same $V_{DS_{SAT}}$ voltage, and that process and temperature
variations and the body effect on $M_{1A/B}$ are related to a threshold voltage variation
of ΔV_T. The minimum $V_{iCM_{DC}}$ can also be written as Eq. (2.8), where V_{OD} is the
overdrive voltage of M1a/b.

$$V_{DD_{min}} = 3.V_{DS_{SAT}} + \Delta V_T \qquad (2.7)$$

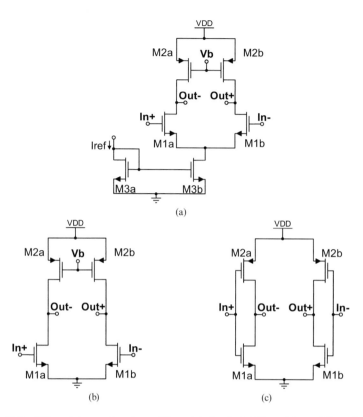

Fig. 2.9 Fully-differential operational amplifiers: (**a**) classical differential amplifier using three-stacked transistors, (**b**) pseudo-differential and (**c**) inverter-based amplifiers with two-stacked transistors

$$V_{iCM_{DC}} = V_{DS_{SAT}} + \Delta V_T + V_{OD} + V_T \qquad (2.8)$$

In the sub-micron and nanometer CMOS processes, the V_T is raging from 150 mV to 400 mV, according to the transistor type, and the ΔV_T is about 20%. The minimum $V_{DS_{SAT}}$ is obtained in weak inversion, and it is about four times the thermal voltage, approximately equal to 100 mV at the room temperature. The overdrive voltage is dependent on the desired inversion level and should be lower than -60 mV to find the WI operation [8]. By Eqs. (2.7) and (2.8) and using $V_T = 250\,\text{mV}$ and $V_{OD} = -100$ mV, the minimum power supply and the minimum input common-mode voltages are equal to 350 mV and 300 mV, respectively. If the circuit has some strategies to compensate the ΔV_T variation, such as the bulk forward bias discussed in Sect. 2.1.3, this part can be removed from the previous equations and the minimum considered values are reduced to about 300 mV and 250 mV, respectively.

In previous equations, the minimum value is achieved by assuming no output swings and operation at room temperature. Thus, practical implementations of the conventional differential amplifier are limited to a minimum supply voltage over 500 mV [2].

The minimum supply voltage can be further reduced if the common-mode current source transistor is removed from the differential amplifier, as shown in the schematic of Fig. 2.9b. Now the amplifier works as a pseudo-differential amplifier, using two common-source amplifiers. As this circuit has no internal nodes, the minimum supply voltage is not dependent on the threshold voltage variation. Then, the minimum V_{DD} voltage is reduced to twice the $V_{DS_{SAT}}$, as given by Eq. (2.9), resulting in the minimum value of 200 mV when in WI. The minimum input DC common-mode voltage becomes equal to the V_{GS} voltage needed to reach the target overdrive voltage in the input transistor, as shown in Eq. (2.10). Additionally, due to the circuit symmetry of the pseudo-differential amplifier, a higher output swing voltage is obtained by using a DC output common-mode voltage of $V_{DD}/2$.

$$V_{DD_{min}} = 2.V_{DS_{SAT}} \tag{2.9}$$

$$V_{iCM_{DC}} = V_{GS_1} = V_{OD} + V_{Th} \tag{2.10}$$

The same voltage limit of the pseudo-differential amplifier is obtained by using the inverter-based fully-differential amplifier shown in Fig. 2.9c. The only difference is due to the $V_{iCM_{DC}}$ that should satisfy both the NMOS and PMOS transistors desired overdrive voltages, making the $V_{iCM_{DC}}$ around the $V_{DD}/2$ level the best bias point.

The $V_{DD_{min}}$ can also be reduced in amplifiers with three or more stacked transistors by using circuit strategies that allow the use of unsaturated transistors without drastic reductions on the voltage gain. Such kind of strategy is employed by [10] to design a 250 mV single-ended amplifiers using the cascode effect to operate with the V_{DS} of only 50 mV in some transistors.

2.2.2   Low Power ULV Operational Amplifier

In contrast to the $V_{DD_{min}}$ voltage that can be expressed as a function of the transistor saturation voltage, the operational amplifier minimum power dissipation is related to the circuit topology characteristics and the circuit specification values required by the application. In the target application of active-RC filters and programmable gain amplifiers, the operational amplifier is classically designed to present a high voltage gain, the capability of driving resistive loads and bandwidth much higher than the maximum filter frequency.

The gain-bandwidth product (GBW) required to the operational amplifiers used in the active-RC filters has the minimum value given as a function of the filter quality factor (Q_{filter}) and the cutoff frequency (f_{cutoff}) as shown in Eq. (2.11) [33].

$$GBW_{min} = 8.Q_{filter} \cdot f_{cutoff} \tag{2.11}$$

The GBW specification of the pseudo-differential amplifier, shown in Fig. 2.9b, is approximately equal to $g_m/(2\pi.C_L)$. Where g_m is the amplifier equivalent transconductance and C_L is the total output load capacitance. Assuming a filter f_{cutoff} of 1 MHz, Q_{filter} of 1.25 and loading capacitance of 1 pF, the minimum GBW required to the amplifier is 10 MHz. To find the target GBW_{min} the pseudo-differential amplifier should present a g_m value of approximately 60 μS. In the weak inversion level, the transistor g_m/I_D ratio is around 30 V^{-1}, that results in a minimum drained current of 2 μA in each branch of the pseudo-differential amplifier and 2 μW of power dissipation if it is powered with a 0.5 V supply. If the inverter-based amplifier, shown in Fig 2.9c, is considered to the same example the minimum power dissipation at 0.5 V is reduced to about 1 μW due to the contribution of both NMOS and PMOS transistors on the equivalent g_m. The minimum power dissipation is never reached by real applications due to the dissipated power on the extra circuits needed to the complete amplifier implementation and to the difficulties of operating at the Mega-Hertz frequency range using weak channel inversion devices.

The voltage gain of a low-voltage amplifier can be improved by using multiple-stage amplifiers. In such amplifiers, two or more stages are cascaded to improve the voltage gain and to present the high input and low output impedances desired to the operational amplifier. However, to have a reasonable open-loop phase margin and make the closed-loop operation stable, some Miller based phase margin compensation or feed-forward techniques should be added to the circuit [27]. The phase margin compensation results in higher power dissipation, as the circuits presented by [2, 5] or in reduced bandwidth, as the circuit presented by [24].

On the other hand, single-stage (unbuffered) amplifiers do not need margin phase compensation and can have higher bandwidth and lower power dissipation, at the same time. However, the maximum voltage gain of these amplifiers is limited to be around 30 dB, reducing the obtained linearity when in closed-loop operation. According to [33], the effects of the low voltage gain of single stage amplifiers can be tolerated by the modern wireless receivers since the design focus is to maximize the bandwidth with reduced power dissipations.

The main challenge of using a single-stage amplifier in the active-RC filters is the effect of the resistive load that reduces, even more, the voltage gain. [25] and [33] proposed the use of a single-stage operational transconductance amplifier (OTA) with output buffers to isolate the resistive feedback load from the amplifier output. However, the topologies of OTAs and buffers employed are not suitable for ULV operation due to the number of stacked saturated transistors. The use of negative conductance/transconductance connected to the amplifier outputs to reduce the resistive and the output conductance loading effects are presented by [6, 32]. Due to the risk of oscillation, the output load cannot be completely canceled, and the circuit implementation can present reduced linearity when operating at the maximum output voltage amplitude. To reduce the linearity issue the authors [17, 30, 34] proposed the use of negative conductances and transconductance connected at the operational amplifier virtual ground, where the voltage swing is smaller. In [30], a pseudo-differential amplifier and a negative input transconductance are used to implement a low power 0.5 V active-RC filter. However, the used pseudo-differential OTA has no common-mode rejection, and the negative transconductance is not compensated for process and temperature variations, that are present in this kind of circuit implementation.

2.2.3   The Common-Mode Rejection of the ULV Amplifiers

The minimum V_{DD} voltage is obtained in ULV amplifier by removing the common-mode current source transistor from the fully-differential amplifiers. The generated pseudo-differential amplifier has no common-mode (CM) rejection since it works as two independent single-ended amplifiers. Therefore, other circuits should be added to the pseudo-differential amplifier in order to reduce the common-mode voltage gain and, consequently, to increase the common-mode rejection rate (CMRR). Additionally to that, the fully-differential amplifiers should have a common-mode feedback (CMFB) circuit to keep the output common-mode voltage equal to a reference DC level ($V_{CM_{ref}}$).

The ULV pseudo-differential amplifier proposed by [5] employed a feedforward common-mode rejection circuit [21], connected in parallel to the input differential pair, and a local CMFB circuit. The CMFB uses CM sense resistors and a controlled DC current source to fed the CM signal back and also to provide the voltage bias to the active load. The main disadvantage of the proposed technique is the increased input capacitance since the amplifier input terminals are connected to both the differential pair and to the feedforward circuit.

The alternative to achieve a reduced input capacitance is by performing both the common-mode rejection and the output common-mode control using the CMFB circuit [13–15, 35]. Grasso et al. [13] proposed a switched capacitor CMFB that uses the bulk terminal of the active load to fed the signal back. Zhang et al. [35] and Khateb and Kulej [16] solved the problem of the conventional differential difference amplifier (DDA) implementation [9] at the ULV range by designing bulk-driven circuits able to operate at 0.6 V and 0.3 V power supply, respectively. Ismail and Mostafa [15] and Harjani and Palani [14] proposed CMFB circuits for inverter-based amplifiers using common-mode sense resistors and a pseudo-differential error amplifier. The CM control is performed by using current source transistors connected in parallel to the main inverter amplifier in order to source/sink current to/from the output nodes, keeping the CM output level constant. The main disadvantage of the common-mode rejection using CMFB-based approaches is the high bandwidth needed at the CMFB loop to provide the CM rejection in whole the amplifier bandwidth.

The common-mode rejection rate of the inverter-based amplifiers can also be improved by using some extra CMOS inverters circuits in the feedforward or feedback modes [22, 31]. In [22] an output to input connected CMOS inverter and a cross-coupled negative transconductor are used at the amplifier output to make the

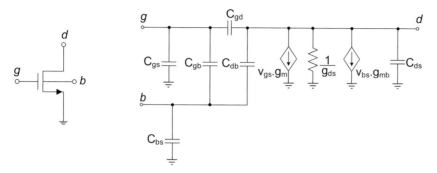

Fig. 2.10 Small-signal model of a NMOS transistor

common-mode voltage gain equal to 0 dB. In [31] a similar approach is employed but the output to input connected CMOS inverters are used to build voltage or current follower circuits. The main advantage of these strategies is the no need for an error amplifier and common-mode sense resistors. However, these circuits are sensitive to PVT variations and have limited linearity at high output swing levels.

2.2.4 Gate and Bulk Input Amplifiers

The CMOS transistors can be modeled for the small-signal operation using a common-source representation, as shown in Fig. 2.10. The circuit representation is composed of the dependent current sources related to the AC V_{gs} and V_{bs} voltages and the gate and bulk transconductances, the output resistance of $1/g_{ds}$ and the parasitic capacitances between each one of the gate, drain, bulk and source nodes.

We can realize through the model representation that both the gate and bulk terminals have similar AC behavior and can be used as input for the voltage amplification. Classically, only gate input operational transconductor amplifier were used in the implementation of the active-RF filters due to the lower bulk noise contribution [19]. However, the bulk forward bias benefits and the improved input common-mode range have motivated the development of several bulk-driven operational amplifier circuits [1, 5, 10, 20, 23, 24, 28, 36].

The main difference between both approaches is related to the value of the input transconductance that is lower at the bulk terminal. The g_{mb}/g_m ratio can be estimated by using Eq. (2.12) which is proportional to the inverse of the V_{BS} voltage [36], and is reduced by the CMOS process scaling [10]. In the literature, a range from 0.15 to 0.25 is presented for the g_{mb}/g_m ratio at the WI and MI levels.

$$g_{mb}/g_m = \frac{\gamma}{2.\sqrt{2.\phi_F + V_{BS}}} \tag{2.12}$$

The bulk-driven amplifier has lower voltage gain and bandwidth in comparison to the gate input version because of the smaller transconductance. Thus, the bulk-driven amplifiers are applied only in low-frequency circuits using multiple stages topologies.

The main advantage of the bulk-driven approaches, at the ULV range, is the improved channel inversion level capability since the gate terminal is independent on the input signal, allowing the use of higher V_{GS} voltages for the transistor bias.

2.2.5 Compensation of the PVT Variation on the ULV Amplifiers

The circuit operation at the reduced power supply voltage naturally increases the circuit sensitivity to the process, voltage and temperature (PVT) variations. It becomes even more significant when combined with pseudo-differential amplifier and inverter-based circuits. In these circuits, the DC output CM voltage (V_{oDC}) and the current drained from the power supply are very sensitive to the PVT variations. The common-source pseudo-differential amplifier, as shown in Fig. 2.9b, has the PVT compensation facilitated because the gate terminal of the current source load can be employed for this propose. In contrast, the inverter-based amplifier, as shown in Fig. 2.9c, has both the NMOS and PMOS gate terminals connected to the amplifier input and cannot be compensated directly.

Several strategies were presented in the literature to design PVT robust ULV circuits. In [5] some replica bias circuits are applied to control both the current and the output voltage of a ULV pseudo-differential amplifier employing the intensive use of bulk forward bias. [31] employed an error amplifier connected to both PMOS and NMOS bulk terminals of the inverter-based OTA to adjust the inverters DC output voltage (trip point). However, the PMOS and NMOS bulk common control cannot work correctly when the process variation tends to the SF or FS corners. The closed-loop compensation circuits presented in [14] are applied to the inverter-based amplifiers where both the V_{oDC} and the drained current are compensated for PVT. The series transistor used to the current compensation reduces the amplifier output swing, and the operation is limited to the power supply range of 0.9 V. In [15] a very efficient approach is proposed to control the V_{oDC} of inverter-based amplifiers by using the CMFB circuit and four common-mode current sources. The V_{oDC} compensation reduces the voltage gain variability, but as the drain current of the main inverter-amplifier is not compensated, the bandwidth is very sensitive to the PVT variation.

In [4] a design strategy using the series-parallel transistor association [11] is proposed to compensate for the mismatch variability and the effects of the Halo implantation on the inverter-based Nauta OTA. As the series association increases the equivalent transistor channel length, the process variability and the Halo implantation effects are reduced. The main advantage of the proposed technique is the no need for extra circuits, however it requires the transistor association, increasing the circuit capacitances and making this approach useful only for low-frequency circuits (Hertz to kilo-Hertz range).

2.3 Conclusion

In this chapter it was shown that the design of ULV and ULP operational amplifiers for active-RC filters have increased complexity in comparison to the conventional topologies.

The CMOS transistor operation, at the ULV range, is limited by the channel inversion levels at the weak and moderated regions. The bulk forward bias can be extensively applied to reduce the V_T, and the transistor channel length should be carefully sized to reduce the reverse short-channel effects.

The use of a single-stage operational amplifier is the key strategy to reduce the circuit power dissipation, but the low voltage gain and the reduced capacity of driving resistive loads impose the need for some extra compensation circuits.

The minimum V_{DD} voltage of the operational amplifiers is reduced by removing the tail common-mode current source of the traditional fully differential amplifier. However, new challenges are imposed to the amplifier design in order to compensate the common-mode gain, the PVT variabilities, and the design of the CMFB circuit, needed at the fully-differential topologies.

References

1. O. Abdelfattah, G.W. Roberts, I. Shih, Y.C. Shih, An ultra-low-voltage CMOS process-insensitive self-biased OTA with rail-to-rail input range. IEEE Trans. Circ. Syst. I: Regul. Pap. **62**(10), 2380–2390 (2015). https://doi.org/10.1109/TCSI.2015.2469011
2. A. Balankutty, S.A. Yu, Y. Feng, P.R. Kinget, A 0.6-V zero-IF/low-IF receiver with integrated fractional-N synthesizer for 2.4-GHz ISM-band applications. IEEE J. Solid-State Circ. **45**(3), 538–553 (2010) https://doi.org/10.1109/JSSC.2009.2039827
3. I.D. Bender, G.S. Cardoso, A.C. De Oliveira, L.C. Severo, A. Girardi, T.R. Balen, Testing fully differential amplifiers using common mode feedback circuit: a case study, in *2015 IEEE 6th Latin American Symposium on Circuits and Systems, LASCAS 2015 - Conference Proceedings*, Montevideo (IEEE, New York, 2015). https://doi.org/10.1109/LASCAS.2015.7250488
4. R.A.S. Braga, H.C. Ferreira, G.D. Colletta, O.O. Dutra, Calibration-Less Nauta OTA Operating at 0.25-V Power Supply in a 130-nm Digital CMOS Process. *2017 IEEE 8th Latin American Symposium on Circuits & Systems (LASCAS 2015)* (2017), pp. 12–15
5. S. Chatterjee, Y. Tsividis, P. Kinget, 0.5-V analog circuit techniques and their application in OTA and filter design. IEEE J. Solid-State Circ. **40**(12), 2373–2387 (2005). https://doi.org/10.1109/JSSC.2005.856280
6. S. Chatterjee, Y. Tsividis, P. Kinget, A 0.5 V filter with PLL-based tuning in 0.18 um CMOS, in *ISSCC. 2005 IEEE International Digest of Technical Papers. Solid-State Circuits Conference, 2005.*, San Francisco (IEEE, New York, 2005), pp. 2004–2006 https://doi.org/10.1109/ISSCC.2005.1494091

7. S. Chatterjee, K.P. Pun, N. Stanić, Y. Tsividis, P. Kinget, *Analog Circuit Design Techniques at 0.5 V* (Springer, New York, 2007)
8. D. Colombo, C. Fayomi, F. Nabki, L.F. Ferreira, G. Wirth, S. Bampi, A design methodology using the inversion coefficient for low-voltage low-power CMOS voltage references. J. Integr. Circ. Syst. **6**(1), 7–17 (2011)
9. J.F. Duque-Carrillo, Control of the common-mode component in CMOS continuous-time fully differential signal processing. Analog Integr. Circ. Signal Process. **4**(2), 131–140 (1993). https://doi.org/10.1007/BF01254864
10. L.H.C. Ferreira, S. Member, S.R. Sonkusale, S. Member, A 60-dB gain OTA operating at 0.25-V power supply in 130-nm digital CMOS process. IEEE Trans. Circ. Syst. I: Regul. Pap. **61**(6), 1609–1617 (2014)
11. C. Galup-Montoro, M.C. Schneider, I.J.B. Loss, Series-parallel association of fet's for high gain and high frequency applications. IEEE J. Solid-State Circ. **29**(9), 1094–1101 (1994). https://doi.org/10.1109/4.309905
12. C. Galup-Montoro, M.C. Schneider, M.B. Machado, Ultra-low-voltage operation of CMOS analog circuits: amplifiers, oscillators, and rectifiers. IEEE Trans. Circ. Syst. II: Express Briefs **59**(12), 932–936 (2012). https://doi.org/10.1109/TCSII.2012.2231042
13. A.D. Grasso, P. Monsurro, S. Pennisi, G. Scotti, A. Trifiletti, Analysis and implementation of a minimum-supply body-biased CMOS differential amplifier cell. IEEE Trans. Very Large Scale Integr. Syst. **17**(2), 172–180 (2009) ISSN:10638210. https://doi.org/10.1109/TVLSI.2008.2003482
14. R. Harjani, R.K. Palani, Design of PVT tolerant inverter based circuits for low supply voltages. Proc. Custom Integr. Circ. Conf. **1**(3) (2015). https://doi.org/10.1109/CICC.2015.7338424
15. A. Ismail, I. Mostafa, A process-tolerant, low-voltage, inverter-based ota for continuous-time $\Delta\Sigma$ ADC. IEEE Trans. Very Large Scale Integr. Syst. **24**(9), 2911–2917 (2016)
16. F. Khateb, T. Kulej, Design and implementation of a 0.3-V differential difference amplifier. IEEE Trans. Circ. Syst. I: Regul. Pap. **66**(2), 513–523 (2019). https://doi.org/10.1109/TCSI.2018.2866179
17. P. Khumsat, A. Worapishet, P. Sirisuk, Single-stage CMOS OTA for active-RC filter design. *2007 Asia-Pacific Conference on Communications, APCC* (2007), pp. 59–62. https://doi.org/10.1109/APCC.2007.4433500
18. T.H. Kim, J. Keane, H. Eom, C.H. Kim, Utilizing reverse short-channel effect for optimal subthreshold circuit design. IEEE Trans. Very Large Scale Integr. Syst. **15**(7), 821–828 (2007). https://doi.org/10.1109/TVLSI.2007.899239
19. P. Kinget, S. Chatterjee, Y. Tsividis, Ultra-low voltage analog design techniques for nanoscale CMOS technologies, in *2005 IEEE Conference on Electron Devices and Solid-State Circuits*, Hong Kong (IEEE, New york, 2005), pp. 9–14. https://doi.org/10.1109/EDSSC.2005.1635192
20. T. Kulej, 0.5-V bulk-driven CMOS operational amplifier. IET Circ. Dev. Syst. **7**(6), 352–360 (2013). https://doi.org/10.1049/iet-cds.2012.0372
21. A.N. Mohieldin, E. Sánchez-Sinencio, J. Silva-Martínez, A fully balanced pseudo-differential OTA with common-mode feedforward and inherent common-mode feedback detector. IEEE J. Solid-State Circ. **38**(4), 663–668 (2003)
22. B. Nauta, A CMOS transconductance-C filter technique for very high frequencies. IEEE J. Solid-State Circ. **27**(2), 142–153 (1992). https://doi.org/10.1109/4.127337
23. S.W. Pan, C.C. Chuang, C.H. Yang, Y.S. Lai, A novel OTA with dual bulk-driven input stage, in *Proceedings - IEEE International Symposium on Circuits and Systems*, Taipei, vol. 1 (IEEE, New York, 2009), pp. 2721–2724. ISBN:978-1-4244-3828-0. https://doi.org/10.1109/ISCAS.2009.5118364
24. Z. Qin, A. Tanaka, N. Takaya, H. Yoshizawa, 0.5-V–70-nW rail-to-rail operational amplifier using a cross-coupled output stage. IEEE Trans. Circ. Syst. II: Express Briefs **63**(11), 1009–1013 (2016). https://doi.org/10.1109/TCSII.2016.2539081
25. A. Rasekh, M. Sharif Bakhtiar, Design of low-power low-area tunable active RC filters. IEEE Trans. Circ. Syst. II: Express Briefs 2017. https://doi.org/10.1109/TCSII.2017.2658635

26. M.C. Schneider, C. Galup-Montoro, *CMOS Analog Design Using All-Region MOSFET Modeling*, 1st edn. (Cambridge University Press, Cambridge, 2010)

27. B.K. Thandri, J. Silva-martínez, S. Member, A robust feedforward compensation scheme for multistage operational transconductance amplifiers with no miller capacitors. IEEE J. Solid-State Circ. **38**(2), 237–243 (2003)

28. M. Trakimas, S. Sonkusale, A 0.5 V bulk-input OTA with improved common-mode feedback for low-frequency filtering applications. Analog Integr. Circ. Signal Process. **59**(1), 83–89 (2009). https://doi.org/10.1007/s10470-008-9236-z

29. Y. Tsividis, *Operation and Modeling of the MOS Transistor*, 2nd edn. (Oxford University Press, New York, 2003). ISBN:978-0-1951-7015-3

30. C. Upathamkuekool, A. Jiraseree-amornkun, J. Mahattanakul, A compensation technique for compact low-voltage low-power active-RC filters. IEEE Int. Symp. Circ. Syst. **1**(3), 3633–3636 (2010). https://doi.org/10.1109/ISCAS.2010.5537782

31. R.G. Vieru, R. Ghinea, Inverter-based ultra low voltage differential amplifiers, in *Proceedings of the International Semiconductor Conference, CAS*, vol. 2 (2011), pp. 343–346. https://doi.org/10.1109/SMICND.2011.6095811

32. J.Y.J. Yan, R. Geiger, A high gain CMOS operational amplifier with negative conductance gain enhancement, in *Proceedings of the IEEE 2002 Custom Integrated Circuits Conference* (2002), pp. 337–340. https://doi.org/10.1109/CICC.2002.1012835

33. L. Ye, C. Shi, H. Liao, R. Huang, Y. Wang, Highly power-efficient active-RC filters with wide bandwidth-range using low-gain push-pull Opamps. IEEE Trans. Circ. Syst. I: Regul. Pap. **60**(1), 95–107 (2013)

34. S. Zeller, C. Muenker, R. Weigel, U. Ussmueller, A 0.039 mm2 inverter-based 1.82 mW 68.6 dB-SNDR 10 MHz-BW CT-$\Sigma\Delta$ -ADC in 65 nm CMOS using power- and area-efficient design techniques. IEEE J. Solid-State Circ. **49**(7), 1548–1560 (2014). https://doi.org/10.1109/JSSC.2014.2321063

35. J. Zhang, Y. Lian, L. Yao, B. Shi, A 0.6-V 82-dB 28.6-uW continuous-time audio delta-sigma modulator. IEEE J. Solid-State Circ. **46**(10), 2326–2335 (2011). ISSN:0018-9200. https://doi.org/10.1109/JSSC.2011.2161212

36. L. Zuo, S.K. Islam, Low-voltage bulk-driven operational amplifier with improved transconductance. IEEE Trans. Circ. Syst. I: Regul. Pap. **60**(8), 2084–2091 (2013) https://doi.org/10.1109/TCSI.2013.2239161

Chapter 3
Single Stage OTA and Negative Transconductance Compensation

3.1 The Use of a Negative Transconductor for Single-Stage OTA Compensation

The filters and programmable gain amplifiers presented in this book are based on the use of an active-RC closed-loop circuits combining a single-stage OTA and a negative input transconductance. The small-signal analysis of the closed-loop amplifier and the integrator are presented in this section. Additionally, the noise contribution of the input negative transconductor is also analyzed.

3.1.1 Closed-Loop Amplifier

A closed-loop fully differential amplifier using a single-stage OTA and a negative input transconductance is shown in Fig. 3.1a. Its differential-mode (DM) small-signal circuit is shown in part b of the same figure. The single-stage OTA was modeled as a single-pole amplifier composed of the transconductance (g_m), the output conductance (g_{ds}) and an output capacitance (C_o). The negative input transconductance (g_{mneg}) is represented by a negative conductance connected at the v_x node.

The closed-loop voltage gain ($Av_{cl} = v_{out}/v_{in}$) of this circuit can be evaluated using Eq. (3.1) and, at low frequencies, it is simplified to Eq. (3.2). The low-frequency closed-loop gain (Av_{cl0}) of this circuit usually is lower than the ideal gain of R_2/R_1 because of the OTA reduced voltage gain and the loading effects. However, by using $g_{mneg} \rightarrow -1/R_1 - 1/R_2$ these effects can be canceled and the ideal gain is reached for any values of R_2/R_1 [12]. In terms of gain, this strategy reaches the same results as those using a high open-loop voltage gain buffered operational amplifier but presenting low power dissipation. Additionally,

© The Author(s), under exclusive license to Springer Nature Switzerland AG 2022
L. C. Severo, W. A. M. Van Noije, *Ultra-low Voltage Low Power Active-RC Filters and Amplifiers for Low Energy RF Receivers*,
https://doi.org/10.1007/978-3-030-90103-5_3

Fig. 3.1 Closed-loop
fully-differential amplifier
using single-stage OTA and a
negative input
transconductance (**a**) and its
small-signal model
considering a single-pole
OTA (**b**) and considering the
input and feedback
capacitances (**c**)

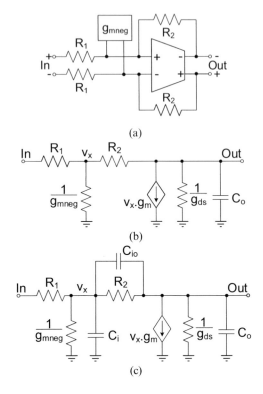

the optimal negative transconductance is not dependent on the OTA parameters
(g_m and g_{ds}), and the compensation works even with very reduced voltage gain
OTAs. At higher frequencies, the closed-loop bandwidth is limited by the OTA pole
frequency (ω_p) evaluated using Eq. (3.3). It is also dependent on the negative input
transconductance, and it tends to infinity when $g_{mneg} \rightarrow -1/R_1 - 1/R_2$. Thus, the
use of $g_{mneg} \rightarrow -1/R_1 - 1/R_2$ could be very important to compensate both the
low voltage gain and the reduced bandwidth of low power OTAs.

$$Av_{cl} = -\frac{R_2}{R_1} \cdot \left\{ \frac{g_m - \frac{1}{R_2}}{g_m - \frac{1}{R_2} + [R_2.(s.C_o + g_{ds}) + 1] \cdot \left(\frac{1}{R_1} + \frac{1}{R_2} + g_{mneg}\right)} \right\}$$
(3.1)

$$Av_{cl_0} = -\frac{R_2}{R_1} \cdot \left[\frac{g_m - \frac{1}{R_2}}{g_m - \frac{1}{R_2} + (R_2.g_{ds} + 1) \cdot \left(\frac{1}{R_1} + \frac{1}{R_2} + g_{mneg}\right)} \right]$$
(3.2)

$$= -\frac{R_2}{R_1}\bigg|_{g_{mneg}=-1/R_1-1/R_2}$$

$$\omega_p = -\frac{1}{R_2.C_o} \cdot \left[\frac{g_m + g_{ds} \cdot \left(\frac{R_2}{R_1} + R_2.g_{mneg} + 1 \right) + \frac{1}{R_1} + g_{mneg}}{\frac{1}{R_1} + \frac{1}{R_2} + g_{mneg}} \right] \qquad (3.3)$$

$$\rightarrow \infty|_{g_{mneg} \rightarrow -1/R_1 - 1/R_2}$$

$$Av_{cl} = -\frac{R_2}{R_1} \cdot \left\{ \frac{-s.C_{io} + g_m - \frac{1}{R_2}}{s^2.R_2.C_I + s.[R_2.C_{io}.(g_m + g_{ds}) + C_i.(R_2.g_{ds} + 1) - C_{io}] + g_m - \frac{1}{R_2}} \right\} \qquad (3.4)$$

$$C_I = C_o.C_{io} + C_i.C_{io} + C_i.C_o \qquad (3.5)$$

Where: R_1 and R_2 are the closed-loop resistances, g_m is the OTA single-stage equivalent transconductance, g_{ds} is the OTA equivalent output conductance and g_{mneg} is the negative input transconductance, C_o is the total output capacitance, C_i is the parasitic input capacitance and C_{io} is the feedback capacitance.

Despite of the good improvements, this compensation scheme starts to oscillate when $g_{mneg} = -1/R_1 - 1/R_2$, making the optimal gain and bandwidth compensation not practical. The analysis presented for single-stage OTAs is very similar to the analysis presented by Boutin in 1981 [2], by using a single-ended single-pole high gain operational amplifier and an ideal negative input converter (NIC). However, these analyses have not considered the real parasitic capacitances present at the input (C_i) and between the output and input (C_{io})—feedback capacitance. The real small-signal model considering these capacitances is shown in Fig. 3.1c. The analysis of this circuit using $g_{mneg} = -1/R_1 - 1/R_2$ results in Eq. (3.4) which can be applied to evaluate the closed-loop gain. Now the circuit presents a right-half-plane zero at the frequency $\omega_z = (g_m - 1/R_2)/C_{io}$ and two poles at frequencies ω_{p1} and ω_{p2}. The pole frequencies are obtained by solving the denominator roots of Eq. (3.4) and, according to the passive devices values and the OTA parameters, ω_{p1} and ω_{p2} can be distinct real roots or complex-conjugate roots.

A closed-loop equivalent single-pole amplifier approximation is obtained when ω_{p2} is much higher than ω_{p1} in order to have ω_{p1} as the dominant pole. In such applications, the input and output capacitances are generally defined by the circuit design and load requirements. Thus only the OTA parameters, R_1, R_2 and the feedback capacitor are the designer free variables. Assuming $g_m = 300\,\mu S$, $g_{ds} = 10\,\mu S$, $R_1 = R_2 = 100\,k\Omega$, $C_i = 0.5\,pF$ and $C_o = 4\,pF$, the closed-loop poles frequencies can be analyzed by using Eqs. (3.4) and (3.5). Figure 3.2a shows the values of ω_{p1} and ω_{p2} when C_{io} is changed from 0.1 to 10 pF. In this example, $C_{io} < 0.8\,pF$ results in complex-conjugate poles while $C_{io} > 0.8\,pF$ results in real and independent poles, and ω_{p2} is higher than $10.\omega_{p1}$ for $C_{io} > 1.5\,pF$. Figure 3.2b shows the frequency response of the closed-loop gain for C_{io} equal to 0.01 pF, 0.25 pF, 0.5 pF and 1 pF. For C_{io} down to 0.01 pF the transfer function has a higher peak that can results in instabilities at that frequency, as previous analyzed.

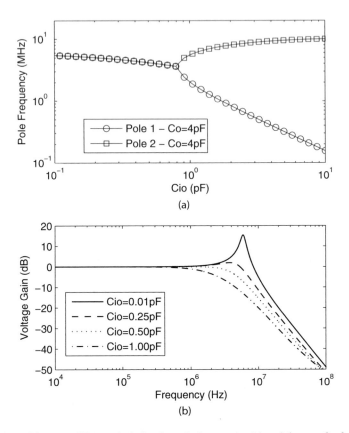

Fig. 3.2 Closed-loop amplifier analysis for the pole frequencies (**a**) and the transfer function (**b**) in relation to the feedback capacitor (C_{io}). Considering $g_{mneg} = -1/R_1 - 1/R_2$, $g_m = 300\,\mu S$, $g_{ds} = 10\,\mu S$, $R_1 = R_2 = 100\,k\Omega$, $C_i = 0.5\,pF$ and $C_o = 4\,pF$

The small-signal analysis presented in this work showed that the closed-loop amplifier is stable for $g_{mneg} = -1/R_1 - 1/R_2$ if higher values of C_{io} are considered. For this values of g_{mneg} the low-frequency gain is completely compensated and is equal to the ideal gain of R_2/R_1, independently of the OTA parameters. The bandwidth compensation does not work as in the ideal single pole OTA, because of the input and feedback capacitors of the real implementations. However, it is not a problem in active-RC filters, the target application of this work, because the feedback capacitor is also used to set the filter cutoff frequency.

The use of a negative input transconductance also can improve the linearity of ULV amplifiers in comparison to a negative output transconductance compensation, as presented in [3], because the amplifier voltage swing at the input is much lower than at the output. Additionally, at the input, the negative transconductor is not dependent on the OTA parameters as it is when connected at the output.

Fig. 3.3 Active integrator using negative input transconductance and single-stage OTA (**a**) and its small-signal representation (**b**)

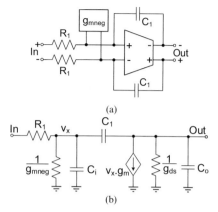

(a)

(b)

3.1.2 Active Integrator

The active integrator is very similar to the closed-loop amplifier and is also very important for active-RC filters. Figure 3.3a shows the schematic of the active integrator using the single-stage OTA and the negative input transconductor. It does not have a DC feedback, as the closed loop amplifier, but a feedback capacitor C_1 is used to set the integrator frequency. The small-signal representation of this circuit is shown in Fig. 3.3b. Here, the input parasitic capacitor (C_i) is included, and the parasitic feedback capacitor C_{io} is considered to be a part of the feedback capacitor C_1. The analysis of this circuit results in Eq. (3.6) that is very different from the $1/(s.R_1.C_1)$ equation expected from an ideal active integrator [11]. Similarly to the closed-loop amplifier, the integrator circuit has a high-frequency right-half-plane zero and two poles. The integrator low-frequency gain ($Av_{int} = v_{out}/v_{in}$) can be evaluated with Eq. (3.8). It is limited by the single-stage OTA voltage gain of g_m/g_{ds} when g_{mneg} is ignored. However, it is entirely compensated by using $g_{mneg} \rightarrow -1/R_1$ that makes the gain tends to infinity, as in the ideal integrator. The use of $g_{mneg} \rightarrow -1/R_1$ also moves the dominant pole frequency to 0Hz while the non-dominant pole is approximately equal to the OTA unity gain frequency of g_m/C_o when $g_m \gg g_{ds}$ and $C_o \gg C_i$, as shown in Eq. (3.9).

$$Av_{int} = -\frac{1}{R_1} \cdot \left\{ \frac{-s.C_1 + g_m}{s^2.C_{II} + s.[C_1.(g_m + g_{ds}) + C_i.g_{ds}] + [g_{ds} + s(C_1 + C_o)].(g_{mneg} + \frac{1}{R_1})} \right\}$$
(3.6)

$$C_{II} = C_o.C_1 + C_i.C_1 + C_i.C_o$$
(3.7)

$$Av_{int_0} = -\frac{1}{R_1} \cdot \left\{ \frac{g_m}{g_{ds} \cdot \left(g_{mneg} + \frac{1}{R_1} \right)} \right\} \rightarrow -\infty|_{g_{mneg} \rightarrow -1/R_1}$$
(3.8)

Fig. 3.4 Circuit used to
perform the output equivalent
noise power analysis of the
closed-loop amplifier without
negative input transconductor
(**a**) and its DM small-signal
circuit (**b**)

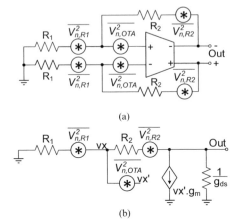

(a)

(b)

$$\omega_{Pint2} = -\left.\frac{C_1.(g_m + g_{ds}) + C_i.g_{ds}}{C_1.(C_o + C_i) + C_i.C_o}\right|_{g_{mneg} \to -1/R_1}$$

$$\approx -\left.\frac{g_m}{C_o}\right|_{g_m \gg g_{ds}, C_o \gg C_i \ and \ g_{mneg} \to -1/R_1} \tag{3.9}$$

The negative input transconductance compensation makes the real integrator tends to the ideal integrator performance, but the circuit is susceptible to phase inversion that can results in instabilities if $g_{mneg} < -1/R_1$ [13]. To avoid instabilities, g_{mneg} should be equal to $-1/R_1 + \Delta_{gm}$, where Δ_{gm} is the safety margin, defined according to the negative input transconductor and the passive devices variability. Additionally, the use of the safety margin helps to increase the input impedance of the active integrator that tends to be very small when g_{mneg} is close to $-1/R_1$.

3.1.3   Noise Analysis

The presented closed-loop compensation technique uses a negative transconductor connected at the single-stage OTA inputs. Because of that, an increase in the equivalent output noise is expected in comparison to the circuit without using the input negative transconductor.

The equivalent noise power at the outputs of the closed-loop single-stage OTA, without the negative input transconductor, can be evaluated by using the circuit presented in Fig. 3.4a. This circuit regards the noise power sources from the OTA and the resistors R_1 and R_2. The OTA noise power source is represented by the equivalent input-referred noise ($V_{n,OTA}^2$) and the resistors are based on the thermal noise power $\overline{V_{n,R_1}^2}$ and $\overline{V_{n,R_2}^2}$ [9]. Figure 3.4b shows the small-signal DM equivalent circuit where the OTA is modeled using an equivalent transconductance (g_m) and an output conductance (g_{ds}), as performed in Sect. 3.1.1. With this circuit, the output noise contribution of each noise source can be evaluated by using the superposition circuit analysis theorem, as show in Eqs. (3.10) to (3.12) for R_1, R_2 and the OTA. The contribution of R_1, Eq. (3.10), is multiplied by the closed-loop DM voltage gain of $\alpha_1.R_2/R_1$, where α_1 is the gain reduction due to the single-stage OTA and it is lower than the unity as shown in Eq. (3.1). The parameters α_2 and α_3 used in Eqs. (3.11) and (3.12) are similar to the parameter α_1. Based on the individual noise contribution, Eq. (3.13) is obtained to the equivalent total noise power at the amplifier output. The multiplication factors of the noise contribution from R_1 and OTA are related to the voltage gain of R_2/R_1, while the contribution of R_2 does not depend on the voltage gain. Assuming a closed-loop voltage gain of $\alpha_1.(R_2/R_1)$, the input-referred noise power of the closed-loop amplifier without compensation can be expressed as shown in Eq. (3.14).

$$\overline{V_{n,out}^2}\Big|_{R_1} = \left(\alpha_1.\frac{R_2}{R_1}\right)^2.\overline{V_{n,R_1}^2} \tag{3.10}$$

$$\overline{V_{n,out}^2}\Big|_{R_2} = \left[\frac{R_1.g_m + 1}{R_1.g_m + 1 + g_{ds}(R_1 + R_2)}\right]^2.\overline{V_{n,R_2}^2} = \alpha_2^2.\overline{V_{n,R_2}^2} \tag{3.11}$$

$$\overline{V_{n,out}^2}\Big|_{OTA} = \left(1 + \frac{R_2}{R_1}\right)^2\left[\frac{R_1.g_m}{R_1.g_m + 1 + g_{ds}.(R_1 + R_2)}\right]^2.\overline{V_{n,OTA}^2}$$

$$= \alpha_3^2.\left(1 + \frac{R_2}{R_1}\right)^2.\overline{V_{n,OTA}^2} \tag{3.12}$$

$$\overline{V_{n,out}^2} = \left(\alpha_1.\frac{R_2}{R_1}\right)^2.\overline{V_{n,R_1}^2} + \alpha_2^2.\overline{V_{n,R_2}^2} + \alpha_3^2.\left(1 + \frac{R_2}{R_1}\right)^2.\overline{V_{n,OTA}^2} \tag{3.13}$$

$$\overline{V_{n,in}^2} = \overline{V_{n,R_1}^2} + \left(\frac{\alpha_2}{\alpha_1}\right)^2\left(\frac{R_1}{R_2}\right)^2.\overline{V_{n,R_2}^2} + \left(\frac{\alpha_3}{\alpha_1}\right)^2.\left(\frac{R_1}{R_2} + 1\right)^2.\overline{V_{n,OTA}^2}$$

$$\approx \overline{V_{n,R_1}^2} + \left(\frac{R_1}{R_2}\right)^2.\overline{V_{n,R_2}^2} + \left(\frac{R_1}{R_2} + 1\right)^2.\overline{V_{n,OTA}^2}\Big|_{\frac{\alpha_2}{\alpha_1} \approx \frac{\alpha_3}{\alpha_1} \approx 1} \tag{3.14}$$

Where: α_1, α_2 and α_3 are reduction gain factors, $\overline{V_{n,R_1}^2}$ and $\overline{V_{n,R_2}^2}$ are the resistor thermal noise power, and $\overline{V_{n,OTA}^2}$ is the OTA input referred noise power.

The resistor thermal noise power is equal to $4.k.T.R$ for a 1 Hz bandwidth [9], where k is the Boltzmann constant, T is temperature (in Kelvin) and R is the resistance value. Replacing $\overline{V_{n,R_1}^2}$ and $\overline{V_{n,R_2}^2}$ by the thermal noise power expression, the simplified input-referred power noise of Eq. (3.14) can be rewritten as Eq. (3.15). With this equation we can conclude that the input-referred noise is only dependent on the noise of the resistor R_1 and the OTA. As expected, the input noise is reduced by increasing the voltage gain (R_2/R_1).

$$\overline{V_{n,in}^2} = \left[4.k.T.R_1 + \overline{V_{n,OTA}^2} \cdot \left(1 + \frac{R_1}{R_2} \right) \right] \cdot \left(1 + \frac{R_1}{R_2} \right) \qquad (3.15)$$

The schematic of the closed-loop amplifier for the noise analysis considering the negative input transconductor is shown in Fig. 3.5a. The negative transconductor noise is represented by the total noise power source $\overline{V_{n,g_{mneg}}^2}$ at the input terminals. The DM small-signal circuit with the negative input transconductor for the noise analysis is shown in Fig. 3.5b. The Thévenin-equivalent circuit, using a noise power source in series with the inverse of the total negative transconductance ($1/g_{mneg}$), is employed to add the negative transconductor noise to the small-signal circuit. By repeating the circuit analysis, using the superposition theorem, the noise contribution of each noise power sources can be evaluated. In these analysis, the optimal value of $g_{mneg} = -1/R_1 - 1/R_2$ is used and because of that, Eq. (3.10) to Eq. (3.12) are rewritten as Eq. (3.16) to Eq. (3.18). Where the α_4 and α_5 are the voltage gain reduction factors and are lower than the unity.

$$\overline{V_{n,out}^2} \Big|_{R_1} = \left(\frac{R_2}{R_1} \right)^2 \cdot \overline{V_{n,R_1}^2} \qquad (3.16)$$

Fig. 3.5 Circuit used to perform the output equivalent noise power analysis of the closed-loop amplifier with negative input transconductor (**a**) and its DM small-signal circuit (**b**)

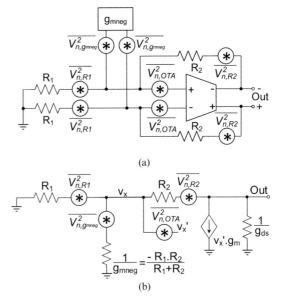

$$\overline{V_{n,out}^2}\bigg|_{R_2} = \left[\frac{g_m.R_2 + 1}{g_m.R_2 + 1 + 2.R_2.g_{ds}}\right]^2 . \overline{V_{n,R_2}^2} = \alpha_4^2 . \overline{V_{n,R_2}^2} \qquad (3.17)$$

$$\overline{V_{n,out}^2}\bigg|_{OTA} = 4.\left[\frac{g_m.R_2}{g_m.R_2 + 1 + 2.R_2.g_{ds}}\right]^2 . \overline{V_{n,OTA}^2} = 4.\alpha_5^2 . \overline{V_{n,OTA}^2} \qquad (3.18)$$

The output noise power contribution due to g_{mneg} is expressed by Eq. (3.19), where $\overline{V_{n,g_{mneg}}^2}$ is multiplied by the non-inverting gain factor of $(R_2/R_1 + 1)$.

$$\overline{V_{n,out}^2}\bigg|_{g_{mneg}} = \left(\frac{R_2}{R_1} + 1\right)^2 . \overline{V_{n,g_{mneg}}^2} \qquad (3.19)$$

Based on Eq. (3.16) to Eq. (3.19), Eq. (3.20) is obtained to the output equivalent noise power of the closed-loop amplifier with the negative input transconductor. By comparing this equation with Eq. (3.13) we can see that the negative transconductor noise contribution has the same multiplication factor as those presented by the OTA in the circuit without compensation. The input-referred noise is obtained by dividing Eq. (3.20) by the closed-loop gain of R_2/R_1 that results in Eq. (3.21). This equation is simplified by disregarding the voltage gain reduction factors, assuming $\alpha_4 \approx \alpha_5 \approx 1$. By replacing the $\overline{V_{n,R_1}^2}$ and $\overline{V_{n,R_2}^2}$ by the thermal noise equation, the simplified input-referred power noise of Eq. (3.21) can be rewritten as Eq. (3.22). Comparing this equation to the input-referred noise without the g_{mneg} it is possible to verify that the negative transconductance noise is added to the input similarly as the OTA input-referred noise. However, the OTA input-referred noise, in the circuit using the g_{mneg} compensation, is reduced by increasing the voltage gain ratio of R_2/R_1 instead of $R_2/R_1 + 1$ as in the circuit without compensation.

$$\overline{V_{n,out}^2} = \left(\frac{R_2}{R_1}\right)^2 . \overline{V_{n,R_1}^2} + \alpha_4^2 . \overline{V_{n,R_2}^2} + 4.\alpha_5^2 . \overline{V_{n,OTA}^2} + \left(\frac{R_2}{R_1} + 1\right)^2 . \overline{V_{n,g_{mneg}}^2}$$
$$(3.20)$$

$$\overline{V_{n,in}^2} = \overline{V_{n,R_1}^2} + \alpha_4^2 . \left(\frac{R_1}{R_2}\right)^2 . \overline{V_{n,R_2}^2} + 4.\alpha_5^2 . \left(\frac{R_1}{R_2}\right)^2 . \overline{V_{n,OTA}^2} + \left(1 + \frac{R_1}{R_2}\right)^2 . \overline{V_{n,g_{mneg}}^2}$$

$$\approx \overline{V_{n,R_1}^2} + \left(\frac{R_1}{R_2}\right)^2 . \overline{V_{n,R_2}^2} + 4.\left(\frac{R_1}{R_2}\right)^2 . \overline{V_{n,OTA}^2} \qquad (3.21)$$

$$+ \left(1 + \frac{R_1}{R_2}\right)^2 . \overline{V_{n,g_{mneg}}^2}\bigg|_{\alpha_4 \approx \alpha_5 \approx =1}$$

$$\overline{V_{n,in}^2} = \left[4.k.T.R_1 + \overline{V_{n,g_{mneg}}^2}.\left(1 + \frac{R_1}{R_2}\right)\right].\left(1 + \frac{R_1}{R_2}\right) + 4.\left(\frac{R_1}{R_2}\right)^2 . \overline{V_{n,OTA}^2}$$
$$(3.22)$$

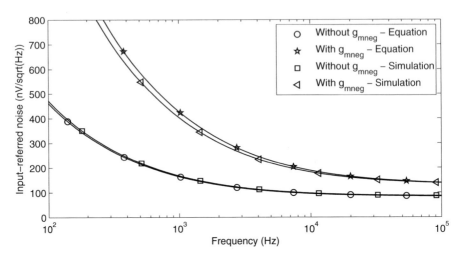

Fig. 3.6 Frequency domain analysis of the closed-loop amplifier input-referred noise with and without the negative input transconductor obtained by using the small-signal equations (3.15) and (3.22) and simulation

Figure 3.6 shows the comparison of the spectral noise density obtained with the simulation and with the use of Eqs. (3.15) and (3.22). For curves based on equations it was employed the simulated values of $\overline{V^2_{n,OTA}}$ and $\overline{V^2_{n,g_{mneg}}}$ related to the frequency. With these curves it is possible to verify the similarity between the calculated and simulated results. The noise addition due to the negative input transconductor can be suppressed in the low energy RF receivers by using a low noise amplifier (LNA) as the first block in the receiver front-end. As given by the Friis equation the IF and the baseband stages noise contributions are reduced by the gain of the preceding stages [10].

3.2 ULV PVT Robust Negative Transconductor

The ULV transconductor presented in this chapter, applied in the single-stage OTA compensation, is based on the classical cross-coupled transconductor shown in Fig. 3.7. The PMOS transistors M1a and M1b are identical and work as cross-coupled transconductances. Their DC drain currents are equal to the reference current (I_{ref}), mirrored from transistor M2c using the current sources transistors M2a and M2b that have the same W/L aspect ratio. This circuit has the small-signal model as shown in Fig. 3.8a. In this circuit g_{m1} is the transconductance of M1a/b, g_{ds1} and g_{ds2} are the output conductances of transistors M1a/b and M2a/b, respectively, C_{gd1} is the parasitic gate to drain capacitance of M1a/b, and C_i is the equivalent parasitic input capacitance. C_i is dependent on the gate to source

Fig. 3.7 Classical cross-coupled negative transconductor

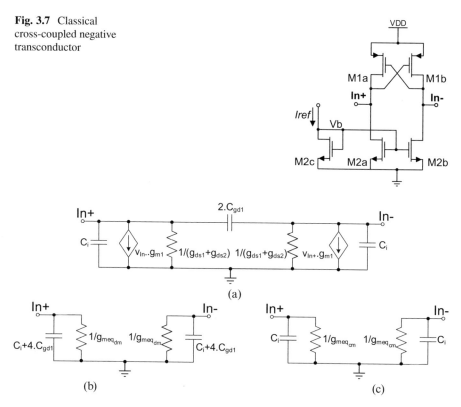

Fig. 3.8 The negative transconductor small-signal model representation: (**a**) complete circuit, (**b**) differential-mode and (**c**) common-mode simplified versions

(C_{gs}), gate to drain (C_{gd}), drain to source (C_{ds}) and drain to bulk (C_{db}) parasitic capacitances of M1a/b and M2a/b, as shown in Eq. (3.23).

$$C_i = C_{gs1} + C_{db1} + C_{gd2} + C_{db2} \tag{3.23}$$

In the differential-mode (DM) the circuit has symmetric input voltages $V_{In+} = -V_{In-}$ and the small-signal model can be simplified to the circuit shown in Fig. 3.8b. The DM equivalent transconductance $(g_{meq_{dm}})$ is negative, as expected, and its value is evaluated with Eq. (3.24). This equation can be rewritten as Eq. (3.25) where the equivalent transconductance is a function of I_{ref} by using the efficiency ratio g_m/I_D and the intrinsic voltage gain g_m/g_{ds} of the NMOS and the PMOS transistors. This equation shows that is possible to adjust the equivalent negative transconductance by changing the I_{ref} current. The DM equivalent parasitic input capacitance is equal to $C_i + 4.C_{gd1}$, where the $4.C_{gd1}$ referenced to ground capacitance is due to the symmetric voltage on the $2.C_{gd1}$ capacitance.

$$g_{meq_{dm}} = -g_{m1} + g_{ds1} + g_{ds2} \tag{3.24}$$

$$g_{meq_{dm}} = -I_{ref} \cdot \left[\left(\frac{g_m}{I_D} \right)_P - \frac{\left(\frac{g_m}{I_D} \right)_P}{\left(\frac{g_m}{g_{ds}} \right)_P} - \frac{\left(\frac{g_m}{I_D} \right)_N}{\left(\frac{g_m}{g_{ds}} \right)_N} \right] \tag{3.25}$$

The common-mode (CM) operation is obtained by using equal input voltages $V_{In+} = V_{In-}$, in which the small-signal circuit can be simplified to the circuit shown in Fig. 3.8c. Now, the equivalent transconductance is given by Eq. (3.26), and it is positive because there is no phase inversion on the cross-coupled transistors. This characteristic is essential in the target application in order to not reduce the closed-loop common-mode rejection and to avoid common-mode instabilities in the active-RC filters. The CM input parasitic capacitance is equal to C_i because there is no CM voltage drop across the $2.C_{gd1}$ capacitor.

$$g_{meq_{cm}} = +g_{m1} + g_{ds1} + g_{ds2} \tag{3.26}$$

Due to the direct relation of the transistor transconductance with the gate to source voltage, the equivalent negative transconductance is very dependent on the voltage at the In+ and In- nodes. Thus, the transconductor is linear only for small DM voltage swings in the In+ and In- nodes. The linearity is not a problem in the presented circuit because the negative transconductance is connected to the input of a closed-loop amplifier, presenting a reduced voltage swing. This is the main advantage of this strategy in comparison to the negative transconductance connected at the OTA outputs, as employed by [3].

The DC voltages at the In+ and In- nodes are also changed by the process and temperature variations, and are equal to the proper gate to source voltage of M1a/b (V_{GS1}) that makes the drain current equal to the mirrored I_{ref} current. As these terminals are connected to the OTA inputs in the active-RC filter, it can shift the OTA input common-mode (CM) voltage, or it can be shifted to $V_{DD}/2$ by the OTA DC control that results in changes on g_{mneg} and DC current flow through the feedback resistors.

To overcome this problem, we have proposed the negative transconductor shown in Fig. 3.9a. A replica bulk forward bias circuit composed of transistors M1r and M2r and an error amplifier is used to compensate the effects of the process and temperature variations. This circuit adjusts the bulk bias voltage (V_{bp}) of M1r to reach $V_{cm} = V_{DD}/2$ when the drain current is equal to the mirrored current I_{ref}. The V_{bp} is also applied to bias the bulk of M1a/b and, as M1r is identical to M1a/b, the DC voltage of nodes In+ and In- ($V_{CM_{DC}}$) also becomes equal to $V_{DD}/2$. The I_{ref} reference current is generated by using a constant g_m bias composed of transistors M2c/d and M7a/b and the external resistor R_{ex}.

Table 3.1 shows the simulated results of the DM equivalent negative transconductance (g_{mneg}) and $V_{CM_{DC}}$ with the replica bulk forward bias and without replica bias (PMOS bulk tied to $V_{DD}/2$) for some process and temperature corners. These results are based on the circuit design for g_{mneg}=28.31 μS at V_{DD}=0.4 V in a 180 nm CMOS process. It is possible to see that the g_{mneg} variations are very close ($\pm$1.85%

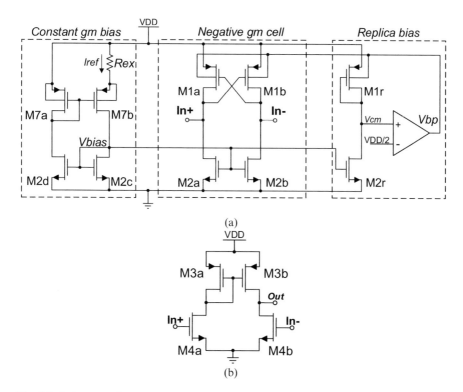

Fig. 3.9 Robust negative transconductor: (**a**) main circuit and (**b**) error amplifier implementation

and $\pm 2.03\%$) for both circuits. However, the $V_{CM_{DC}}$ variation range was reduced from 61.1 mV to less than 7 mV by using the replica bias. This is equivalent to a reduction from $\pm 15.3\%$ to $\pm 1.73\%$.

Additionally, by using the bias circuit, the value of g_{mneg} can be adjusted by changing the reference current (I_{ref}) to a reasonable range, presenting a small DC voltage variation. The variation of the negative transconductance and the DC common-mode voltage as functions of I_{ref} are shown in Fig. 3.10a and b. The g_{mneg} value has the same variation effect with or without the use of replica bias but the $V_{CM_{DC}}$ is kept in a value near to $V_{DD}/2$ for a larger range. The I_{ref} adjustment can be used to compensate for the effect of the resistors and the transconductor variability after the fabrication. Furthermore, the use of the bulk forward bias on the PMOS transistor reduces the threshold voltage (V_{T_p}) by about 15%, increasing the transistor inversion level with no increases in the gate to source voltage.

The presented transconductor uses the simple single-ended pseudo-differential amplifier shown in Fig. 3.9b [7] as error amplifier. It can be designed to dissipate a small quantity of power because it is only used for DC compensation and the loop phase margin is improved by reducing the error amplifier bandwidth.

Table 3.1 Simulation results of the DM equivalent negative transconductance g_{mneg} and V_{DC} input voltage, for some process and temperature corners at $V_{DD}=0.4$ V

	TT	FS@	SF@	FF@	SS@		
Parameter	27 °C	27 °C	27 °C	−40 °C	100 °C	Δ	$\pm\Delta/2$ [%]
$\|g_{mneg}\|$ (no rep. bias) [μS]	28.31	28.17	28.50	28.91	27.86	1.05	±1.85%
$\|g_{mneg}\|$ (with rep. bias) [μS]	28.31	28.41	28.30	28.96	27.81	1.15	±2.03%
$V_{CM_{DC}}$ (no rep. bias) [mV]	200.0	230.9	169.8	210.4	191.7	61.1	±15.30%
$V_{CM_{DC}}$ (with rep. bias) [mV]	200.0	202.5	195.6	201.4	198.7	6.9	±1.73%

Reprinted, with permission, from [5]

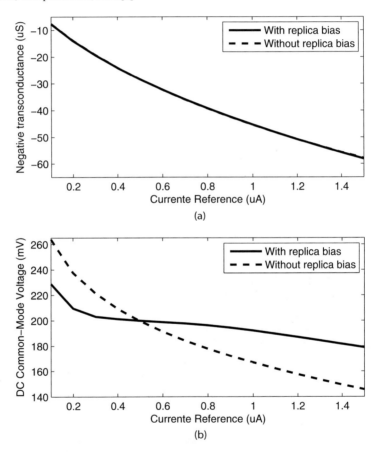

Fig. 3.10 Simulation results of the negative transconductance (**a**), and the DC CM voltage (**b**), versus the reference current with and without the replica bias

As previously analyzed in Sect. 3.1.3, the main drawback of the compensation by using the negative input transconductance is its noise contribution to the closed-loop amplifier equivalent output noise. Based on the CM small-signal circuit of Fig. 3.8, and assuming a thermal noise current of $4.k.T.\gamma_n.g_m$ and an equivalent transconductance given by Eq. (3.26), the total thermal noise power generated by the transconductor can be expressed as Eq. (3.27) [9]. The total noise power can be reduced by increasing the M1a/b transconductance (g_{m1}) and by reducing the transconductance of the bias transistor M2a/b (g_{m2}). However, g_{m1} is dependent on the resistor values used in the closed-loop amplifier. For the optimal gain compensation g_{m1} should be approximately equal to $1/R_1 + 1/R_2$ if $g_{m1} \gg g_{ds1} + g_{ds2}$. Based on this assumption and using $g_{m1} = g_{m2}$ for the sake of simplicity, Eq. (3.27) can be rewritten as Eq. (3.28). Using $R_2 = R_1.Av_{cl} = R$ we can conclude that the only strategy available to reduce the noise contribution is the reduction of the values for the resistors. This conclusion generates a design trade-off between the noise contribution and the power dissipation since for low resistances values the negative transconductor should drain more current from V_{DD} in order to provide more transconductance.

$$\overline{V_{n,g_{mneg}}^2} = 4.k.T.\gamma_n. \left[\frac{g_{m1} + g_{m2}}{(g_{m1} + g_{ds1} + g_{ds2})^2} \right] \approx \frac{4.k.T.\gamma_n}{g_{m1}}. \left(1 + \frac{g_{m2}}{g_{m1}} \right) \Bigg|_{g_{m1} \gg (g_{ds1} + g_{ds2})} \tag{3.27}$$

$$\overline{V_{n,g_{mneg}}^2} \approx 8.k.T.\gamma_n. \left(\frac{R_1.R_2}{R_1 + R_2} \right) \approx 8.k.T.\gamma_n.R. \left(\frac{Av_{cl}}{Av_{cl} + 1} \right) \tag{3.28}$$

Where: γ_n is the transistor thermal noise parameter, k the Boltzmann constant and T is the temperature (in Kelvin).

3.3 ULV Inverter-Based OTA

The OTA presented in this work is based on the use of CMOS inverters circuit to find a high equivalent transconductance to the current ratio (g_{meq}/I_D) and to present a reduced power dissipation. The simplified schematic of the OTA is shown in Fig. 3.11. The circuit has only two-stacked transistors to address the ULV operation and to increase the output voltage swing. As the two CMOS inverter composed of transistors M5a/b and M6a/b are independent of each other, the OTA has the common-mode gain equal to the difference-mode gain. The common-mode rejection rate (CMRR) is improved by using a bulk-driven common-mode feedback (CMFB) approach. The CMFB circuit is also applied to keep the output DC voltage at $V_{DD}/2$ to maximize the output voltage swing. It uses two resistors (Rcm_a and Rcm_b) to measure the output common-mode voltage (V_{OCM}) and an error amplifier to amplify the difference between the measured V_{OCM} and the reference level of $V_{DD}/2$. The CMFB loop is closed by connecting the output of the error amplifier to the PMOS bulk terminal. Due to the bulk forward bias, the threshold voltage of the PMOS transistor is lowered, improving its channel inversion level without changing the gate to source voltage.

Fig. 3.11 Schematic of the ULV OTA—simplified version

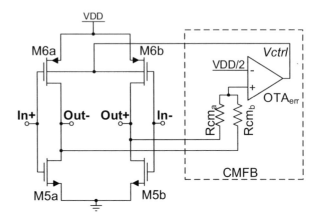

The complete small-signal model of the amplifier is shown in Fig. 3.12a. In the circuit, g_{m5} and g_{m6} are the transconductances of M5a/b and M6a/b, g_{ds5} and g_{ds6} are the output conductances of M5a/b and M6a/b, g_{mb6} is the bulk transconductance of M6a/b, g_{merr} and g_{dserr} are the transconductance and output conductance of the error amplifier, and C_i, C_{io}, C_o, C_{ob}, C_{ierr} and C_{oerr} are the parasitic capacitances. The small-signal circuit can be simplified in the differential-mode operation by removing the CMFB circuit, as shown in Fig. 3.12b. Based on the simplified circuit, the OTA differential-mode voltage gain (Av_{dm}) can be evaluated by Eq. (3.29). At low frequencies, the voltage gain becomes equal to Eq. (3.30) that is dependent on the main CMOS inverters and the resistive load due to the common-mode sense resistors. The circuit frequency response has a single right-half-plane zero and a single pole that can be evaluated using Eqs. (3.31) and (3.32), respectively. Using Eqs. (3.30) and (3.32) the OTA gain-bandwidth product (ω_{GBW})—unity gain frequency - can be evaluated by Eq. (3.33). As the amplifier has a single pole and, generally, the zero is at a very high frequency, the OTA phase margin is typically higher than 60° without the need of any compensation circuit.

$$Av_{dm} = \frac{-s.C_{io} + g_{m5} + g_{m6}}{s.(C_{io} + C_o + C_{ob}) + g_{ds5} + g_{ds6} + \frac{1}{R_{cm}}} \tag{3.29}$$

$$Av_{dm0} = \frac{g_{m5} + g_{m6}}{g_{ds5} + g_{ds6} + \frac{1}{R_{cm}}} \tag{3.30}$$

$$\omega_z = \frac{g_{m5} + g_{m6}}{C_{io}} \tag{3.31}$$

$$\omega_p = -\frac{g_{ds5} + g_{ds6} + \frac{1}{R_{cm}}}{C_{io} + C_o + C_{ob}} \tag{3.32}$$

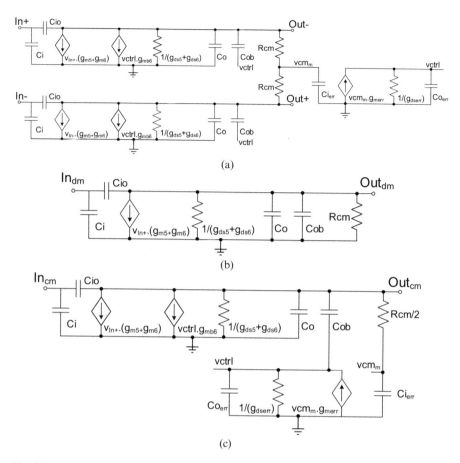

Fig. 3.12 Small-signal circuit of the ULV OTA from Fig. 3.11: (**a**) complete circuit, (**b**) DM and (**c**) CM simplified circuits

$$\omega_{GBW} = \frac{g_{m5} + g_{m6}}{C_{io} + C_o + C_{ob}} \tag{3.33}$$

The common-mode analysis of the small-signal circuit of Fig. 3.12 is performed by changing the CMOS inverter by a single common-mode circuit, as shown in Fig. 3.12c. At low frequency, the common-mode gain can be evaluated using Eq. (3.34). In this equation Av_{err} is the voltage gain of the error amplifier that is equal to g_{merr}/g_{dserr}. The $Av_{err}.g_{mb6}$ product is generated by the CMFB loop and add the common-mode rejection to the ULV OTA. With Av_{dm0} and Av_{cm0} we can evaluate the common-mode rejection rate (CMRR), as shown in Eq. (3.35). The higher the $Av_{err}.g_{mb6}$ product is, the higher the CMRR of the OTA will be.

$$Av_{cm0} = \frac{-(g_{m5} + g_{m6})}{g_{ds5} + g_{ds6} + Av_{err} \cdot g_{mb6}} \tag{3.34}$$

$$CMRR = \frac{Av_{dm0}}{Av_{cm0}} = \frac{g_{ds5} + g_{ds6} + Av_{err} \cdot g_{mb6}}{g_{ds5} + g_{ds6} + \frac{1}{Rcm}} \approx 1 + \frac{Av_{err} \cdot g_{mb6}}{g_{ds5} + g_{ds6} + \frac{1}{Rcm}} \tag{3.35}$$

The frequency response of the CMFB loop is similar to a three poles amplifier transfer function, resulting in a reduced bandwidth and phase margin. Because of that, the CMRR is not kept constant in all the OTA bandwidth. The first pole is generated by the error amplifier pole that can be controlled by the error amplifier bandwidth. Its frequency can be estimated using Eq. (3.36). The second pole is generated at the output of the CMOS inverters, and it is dependent on the amplifier output conductance and the equivalent output capacitance, as shown in Eq. (3.37). The third pole is generated by the common-mode sense resistors and the parasitic capacitance at the error amplifier input that can be evaluated by using Eq. (3.38). Due to the three poles in the CMFB loop, the $Av_{err} \cdot g_{mb6}$ product is reduced when the frequency increases. Because of that, the common-mode gain of Eq. (3.34) is increased, becoming approximately equal to the absolute value of Av_{dm0} when the product $Av_{err} \cdot g_{mb6}$ is much lower than $g_{ds5} + g_{ds6}$. In other words, the bandwidth of the common-mode rejection in the ULV OTA is very dependent on the bandwidth of the CMFB loop.

$$\omega_{pcm1} = -\frac{g_{dserr}}{C_{oerr} + C_{ob}} \tag{3.36}$$

$$\omega_{pcm2} = -\frac{g_{d5} + g_{d6}}{C_o + C_{ob}} \tag{3.37}$$

$$\omega_{pcm3} = -\frac{1}{Rcm.C_{ierr}} \tag{3.38}$$

Figure 3.13 shows the frequency response of the differential-mode and common-mode voltage gains of the ULV OTA, using typical values for the small signal characteristics and parasitic capacitances present on the implementations shown in Chap. 5. In this design the OTA has low-frequency DM and CM gains of 26.0 dB and −12.1 dB, respectively, that results in a CMRR of 38.1 dB. The common-mode gain remains equal to −12.1 dB for frequencies lower than 500 kHz, where is located the dominant pole of the CMFB loop. For higher frequencies, the common-mode gain is increased and it becomes equal to the differential mode gain at frequencies higher than 25 MHz. The peak of the common-mode gain transfer function can generate a common-mode instability in the closed-loop amplifier. Because of that, the peak region should be larger than the cutoff frequency in the active-RC filter application, as will be detailed in the designs of Sect. 5.

The CMFB also reduces the common-mode voltage gain due to the power supply ($Av_{V_{DD}}$ and $Av_{V_{SS}}$). It is important to increase the OTA power supply rejection

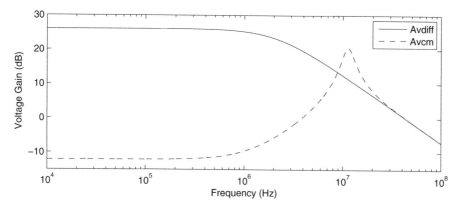

Fig. 3.13 Differential-mode and common-mode gains of the proposed OTA, using typical values for the small signal characteristics and parasitic capacitances

ratios (PSRR) equal to $Av_{dm}/Av_{V_{DD}}$ and $Av_{dm}/Av_{V_{SS}}$. By solving the small-signal circuit of the OTA shown in Fig. 3.11, considering an input AC signal connected to the V_{DD} or V_{SS} (showed as ground) with the inputs tied to the ground, Eqs. (3.39) and (3.40) are obtained for $Av_{V_{DD}}$ and $Av_{V_{SS}}$ at low frequency. These equations have the same gain rejection factor of $Av_{err}.g_{mb6}$ that should be improved to increase PSRR of the OTA. The power supply rejection works in approximately the same bandwidth of the common-mode gain rejection because of the CMFB bandwidth limitation.

$$Av_{V_{DD}} = \frac{g_{m6} + g_{mb6} + g_{ds6}}{g_{ds5} + g_{ds6} + Av_{err}.g_{mb6}} \qquad (3.39)$$

$$Av_{V_{SS}} = \frac{g_{m5} + g_{mb5} + g_{ds5}}{g_{ds5} + g_{ds6} + Av_{err}.g_{mb6}} \qquad (3.40)$$

3.3.1 Improvements in the CMFB Loop

The common-mode rejection of the OTA presented in the last section is very dependent on the CMFB loop. As the common-mode rejection is proportional to the $Av_{err}.g_{mb6}$ product, it can be improved by increasing the error amplifier voltage gain or the bulk transconductance.

The gain of the error amplifier can be increased by using a multiple stage single-ended amplifier or some technique that uses the output conductance cancellation to increase the amplifier output impedance [8]. However, due to the gain bandwidth product, the Av_{err} increase will result in bandwidth decreasing. The bandwidth increase can be obtained by adding more power dissipation to the error amplifier, but

Fig. 3.14 Schematic of the
ULV OTA—improved
version, with the V_{ctrl} voltage
connected to both NMOS and
PMOS bulk terminals

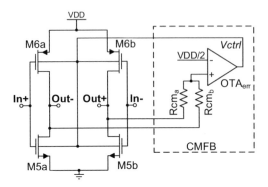

it will reduce the CMFB phase margin because the error amplifier pole is generally the dominant pole of the CMFB loop.

The bulk transconductance can be increased by increasing the PMOS transistor current, but it will also increase the OTA transconductance and output conductance. Another alternative of improvement is by connecting the error amplifier output also to the bulk terminal of the NMOS transistors used in the CMOS inverters, as shown in the schematic of Fig. 3.14. Thus, both PMOS and NMOS transistors of the inverter will be connected to the $Vctrl$ voltage and the equivalent common-mode transconductance (g_{mcm}) will be equal to $g_{mb5} + g_{mb6}$, without any current increase in the OTA. This approach is very efficient and it increases to approximately twice the CM rejection. Furthermore, it reduces the NMOS threshold voltage and also increases the capability of controlling the output DC voltage because the $Vctrl$ DC voltage will have opposite effects on the PMOS and NMOS threshold voltages. The disadvantage of this strategy is the need of a triple-well or buried-N-well CMOS process to provide insulated bulk NMOS transistors. As analyzed in Sect. 2.1.3, the triple-well or buried-N-well transistors are larger than the conventional transistors and more masks are needed in the process fabrication. Additionally, in the modern CMOS processes, the NMOS insulated bulk transistor is commonly available only for the standard V_T transistors.

Figure 3.15 show the comparison of the transfer functions of Fig. 3.13 with the two solutions analyzed, using the typical value of transconductances, output conductances and parasitic capacitances used in the graph of Fig. 3.13. We can observe that increasing twice the Av_{err} the Av_{cm} gain is reduced by approximately 6 dB but its bandwidth is halved. In the other way, by increasing the bulk transconductance twice the same reduction of 6 dB in Av_{cm} is obtained without change the bandwidth. Thus, this comparison showed that is better to increase the g_{mb} instead of increasing the Av_{err} to improve the common mode rejection, keeping the bandwidth without increasing the power dissipation.

In Fig. 3.15 is also shown the Av_{cm} curve using an increase of 10 times in g_{mb}, that results in a reduction of 20 dB in the common-mode gain without changing the bandwidth. However, the circuit of Fig. 3.14 presents the bulk transconductance limited to increase up to about two times. To further increase the equivalent

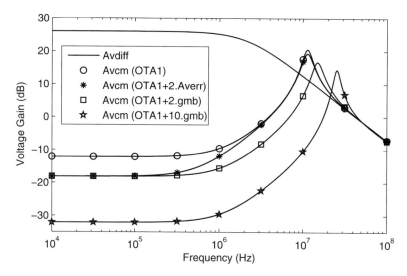

Fig. 3.15 Comparison of the common-mode gain reduction by increasing the error amplifier gain or the bulk transconductance

bulk transconductance (g_{mbeq}) we can use the strategies proposed in [4] and [7] that uses common-mode parallel transistors connected to the output nodes. As was analyzed in Sect. 2.2.4 the small-signal model of parallel transistors has the same behavior as the gate and bulk transconductances, as employed in the previous circuit. Based on that we have presented the circuit shown in Fig. 3.16 by adding transistors M5c, M5d, M6c and M6d with the drain and source terminals connected in parallel to the main inverters and the common gate and the PMOS bulk connected to the $Vctrl$ node. The added transistors work as four current sources to source or sink current from or to the output nodes [7]. In the small-signal model, these transistors work as common-mode transconductors that are parallel to the bulk transconductance, resulting in the equivalent common-mode transconductance given by Eq. (3.41). Now the equivalent transconductance can be increased independently of the main inverter OTA and without using the NMOS bulk. However, due to the parallel transistors, the equivalent output conductance is increased, as shown in Eq. (3.42), and the current sources power dissipation is added to the total OTA power consumption. The increasing in g_{dseq} generates a small reduction in the differential and common-mode gains, as shown in Eqs. (3.43) and (3.44). Thus, there are some trade-offs in the design of this circuit by considering the common-mode rejection, the power consumption and differential-mode gain reduction.

$$g_{mcm} = g_{mb6_{ab}} + g_{mb6_{cd}} + g_{m5_{cd}} + g_{m6_{cd}} \qquad (3.41)$$

$$g_{dseq} = g_{ds6_{ab}} + g_{ds6_{cd}} + g_{ds5_{ab}} + g_{ds5_{cd}} \qquad (3.42)$$

Fig. 3.16 Schematic of the
ULV OTA using parallel
common-mode current
sources

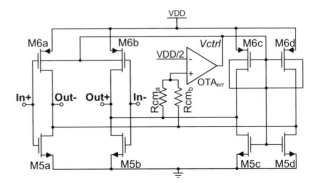

$$Av_{dm0} = \frac{g_{m5_{ab}} + g_{m6_{ab}}}{g_{dseq} + \frac{1}{R_{cm}}} \qquad (3.43)$$

$$Av_{cm0} = \frac{-\left(g_{m5_{ab}} + g_{m6_{ab}}\right)}{g_{dseq} + Av_{err} \cdot g_{mcm}} \qquad (3.44)$$

3.3.2 Improving the Drain Current Control

The OTAs topology presented in previous sections have the DC output common-mode voltage ($V_{o_{CM}}$) controlled by the CMFB circuit in order to be equal to $V_{DD}/2$ even under PVT variations. However, the current drained from the power supply is not controlled and its value is very dependent on the PVT variations. Besides the power dissipation variability, some of the OTA specifications, such as slew rate and unity gain frequency, are directly affected by the current.

In general, the solutions available on the literature able to compensate both the output common-mode DC voltage and the drained current variabilities uses two series transistors at the V_{DD} and GND nodes of the inverter [6]. Such kind of strategy is important because it does not use the bulk terminal of the transistor, but due to the use of four-stacked transistors the OTA output swing and its dynamic range would be reduced.

Due to this issue, we have proposed a solution to further improve the OTA presented in previous sections without using series transistors. The proposed technique uses an independent bulk forward bias voltage to control the NMOS and the PMOS transistors. In the circuit the NMOS bulk bias makes the NMOS V_{DS} voltage equals to $V_{DD}/2$ only if the NMOS current is equal to the target value. Thus, the PMOS bulk bias is forced to bias the circuit in a certain level that makes its current also equals to the target value in order to find DC $V_{oCM} = V_{DD}/2$.

The proposed circuit is shown in Fig. 3.17, which is composed of the same inverter-based OTA presented in the last section, but now using a bulk forward bias

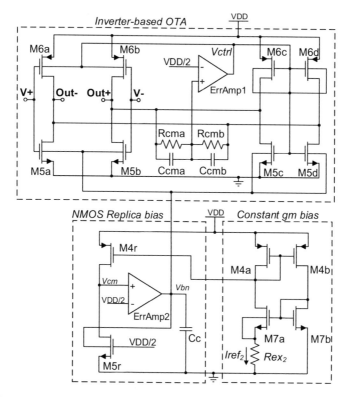

Fig. 3.17 Inverter-based OTA implementation considering the NMOS bulk bias

scheme applied to the NMOS transistors. The NMOS bulk bias was designed with a similar strategy applied to implement the negative transconductor. It is composed of an NMOS replica bias and a constant gm circuit.

The replica bias is composed of the transistors M4r and M5r and the error amplifier 2 (ErrAmp2). The bulk voltage of the transistor $M5r$ (vbn) is adjusted by ErrAmp2 to make the voltage V_{cm} equal to $V_{DD}/2$. Transistors M4a/b and M7a/b together with the external resistor Rex_2 provide a current reference (I_{ref_2}) that is mirrored to M5r by means of M4r. As a consequence of this compensation, transistor M5r has both the V_{DS} voltage and the drain current adjusted to $V_{DD}/2$ and I_{ref_2}, respectively. The $M5r$ bulk voltage (Vbn) is also connected to all the NMOS transistor of the main inverter-based OTA. As these transistors have the same aspect to ratio and V_{GS} voltage of $M5r$, both the drain current and $V_{CM_{out}}$ are compensated to present the desired values under a reasonable range for PVT variations. In other words, the V_{DS} voltage of M5a to M5d will be equal to $V_{DD}/2$ only if the drain current is equal to I_{ref_2}, forcing the CMFB to put a proper DC PMOS bulk voltage to make the drain current of M6a to M6d also equal to I_{ref_2}. The I_{ref_2} value can be calibrated after the fabrication by changing the value of the external resistor Rex_2.

The drain current of transistors M5a to M5d and M6a to M6d can be scaled to $M.I_{ref_2}$ by using an association of M transistors in parallel without changing the compensation capability. As the NMOS bulk control is used only for DC compensation, it can be designed using a very low current in order to maintain the low power dissipation of the OTA.

Additionally, we have improved the CMFB loop phase margin compensation by using the Ccma and Ccmb capacitors in parallel with the common-mode sense resistors in order to reduce the effect of the ErrAmp1 input parasitic capacitance [1].

Figure 3.18 shows some histograms of the CMOS inverter current, the gain-bandwidth product (GBW) and the DC output common-mode voltage of the proposed OTA with and without using the NMOS bulk control. These curves are obtained with a 0.4 V OTA designed in a 130 nm BiCMOS process that is detailed in Chap. 5. The histograms were obtained using Monte Carlo simulations with 1000 samples, considering process and mismatch parameter variation. As can be seen in the histograms the variability of both the inverter current and the GBW are very reduced by using the NMOS replica bias. The DC output common-mode voltage has approximately the same variability, showing that the proposed NMOS bulk bias does not affect the CMFB control.

The current drained from V_{DD} is not as constant, as the inverter current, due to the CM current sources of the CMFB that are changed to adjust the output DC voltage and could be increased or reduced due to the PVT variation. However, the OTA specification will present a reduced variability because they are more sensitive to the main inverter current than to the common-mode current sources. A scheme similar to the OTA presented in Fig. 3.11 can present a very reduced current variability when the NMOS replica bias is applied because since it is dependent only on the main inverter. However, the common-mode rejection cannot be improved twice by using both NMOS and PMOS bulk transconductances because the NMOS bulk terminal is now used for the current control.

3.3.3 Error Amplifier

The error amplifiers employed in the OTA implementation were designed using the same topology of the pseudo-differential single-ended OTA applied in the negative transconductance implementation. However, as the CMFB loop should have a higher bandwidth in comparison to the DC control, we have used a bulk forward voltage in both NMOS and PMOS transistors to find higher channel inversion levels. The schematic of the ErrAmp1 with both PMOS and NMOS bulk tied to $V_{DD}/2$ is shown in Fig. 3.19a. The DC bulk voltage of $V_{DD}/2$ was chosen for M8a/b and M9a/b to have the same aspect ratio of M6a/b and M5a/b from the OTA in order to improve the layout regularity. The schematic of ErrAmp2 is shown in Fig. 3.19b and it was designed as in the negative transconductor to present reduced bandwidth and power dissipation.

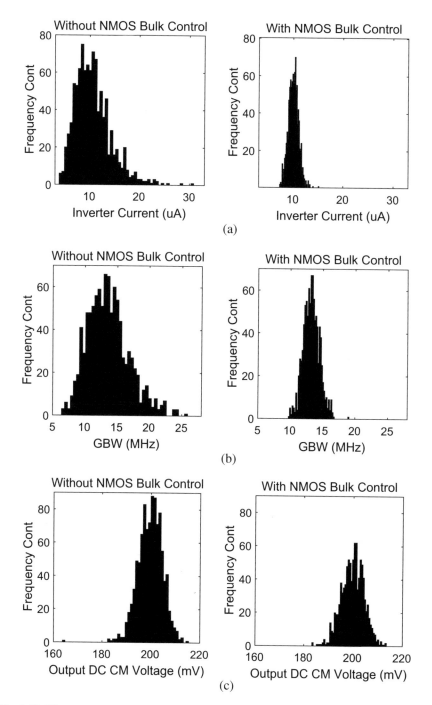

Fig. 3.18 Histogram of some OTA specifications without and with considering the NMOS bulk control: (**a**) CMOS inverter current, (**b**) GBW and (**c**) common-mode output DC voltage

Fig. 3.19 The CMFB error
amplifiers: (**a**) ErrAmp1 and
(**b**) ErrAmp2

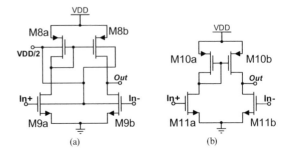

(a)   (b)

3.4   Conclusion

The analysis performed in this chapter were based on the small-signal circuit since the amplifier is used in closed-loop configuration and some compensation circuits to work with very controlled bias voltages are also employed.

The use of the negative input transconductance is very important to compensate for the low voltage gain and the load effect of the single-stage OTAs. The closed-loop voltage gain can be entirely compensated by the input negative transconductor circuit and it is stable if a feedback capacitor is considered. The closed-loop amplifier input-referred noise is increased by the input negative transconductor and its contribution can be reduced by increasing the power dissipation. However, the added noise tends not to be a negative point in the target application of low energy RF receivers because the noise contribution at the IF and baseband stages have a reduced influence on the receiver equivalent noise figure.

A robust negative transconductor, able to operate connected at the OTA inputs, was presented in this chapter. The use of the proposed replica bulk-driven bias controls the DC voltage of the negative transconductance and makes the circuit able to operate in a wide range of adjustable transconductance. The proposed circuit uses bulk forward bias and only two-stacked transistors that make the circuit suitable for ULV operation.

Three versions of ULV inverter-based OTAs were analyzed in this chapter. These circuits use the CMFB loop to add a common-mode rejection to the inverter-based OTA and, consequently, to increase the CMRR and PSRR specifications. The CMFB is also used to make the output DC voltage equal to $V_{DD}/2$ and to maximize the output voltage swing. A novel strategy using the NMOS forward bulk bias was also proposed in this chapter to reduce the OTA current variability without using series transistors. Based on that, it is possible to obtain a two-stacked transistor ULV OTA that presents common-mode rejection and low variability in both the drain current and the DC output voltage.

References

1. R.J. Baker, *CMOS: Circuit Design, Layout, and Simulation*, 3rd edn. (Wiley, Hoboken, NJ, 2011). https://doi.org/10.1002/9780470891179
2. N. Boutin, Active compensation of op-amp inverting amplifier using NIC. Electron. Lett. **17**(25), 978–979 (1981). https://doi.org/10.1049/el:19810682
3. S. Chatterjee, Y. Tsividis, P. Kinget, 0.5-V analog circuit techniques and their application in OTA and filter design. IEEE J. Solid-State Circ. **40**(12), 2373–2387 (2005). https://doi.org/10.1109/JSSC.2005.856280
4. S. Chatterjee, K.P. Pun, N. Stanić, Y. Tsividis, P. Kinget, *Analog Circuit Design Techniques at 0.5 V* (Springer, New York, 2007)
5. L. Compassi-Severo, W. Van Noije, A 0.4-V 10.9-μW/Pole third-order complex BPF for low energy RF receivers. IEEE Trans. Circ. Syst. I: Regul. Pap. **66**(6), 2017–2026 (2019). ISSN:15580806. https://doi.org/10.1109/TCSI.2019.2906206
6. R. Harjani, R.K. Palani, Design of PVT tolerant inverter based circuits for low supply voltages, in *Proceedings of the Custom Integrated Circuits Conference*, vol. 1(3) (2015). https://doi.org/10.1109/CICC.2015.7338424.
7. A. Ismail, I. Mostafa, A process-tolerant, low-voltage, inverter-based OTA for continuous-time $\Delta\Sigma$ ADC. IEEE Trans. Very Large Scale Integr. Syst. **24**(9), 2911–2917 (2016)
8. P. Kinget, S. Chatterjee, Y. Tsividis, Ultra-low voltage analog design techniques for nanoscale CMOS technologies, in *2005 IEEE Conference on Electron Devices and Solid-State Circuits*, Hong Kong, 2005 (IEEE, New York, 2005), pp. 9–14. https://doi.org/10.1109/EDSSC.2005.1635192
9. B. Razavi, *Design of Analog CMOS Integrated Circuits*, vol. 6 (McGraw-Hill, New York, 2001) https://doi.org/10.1111/j.1151-2916.1994.tb07040.x.
10. B. Razavi, *RF Microelectronics*, vol. 53, 2nd edn. (Prentice Hall, New York, 2012). https://doi.org/10.1017/CBO9781107415324.004
11. A.S. Sedra, K.C. Smith, *Microeletrônica* (Pearson, São Paulo, 2007)
12. L. Severo, W. Van Noije, 0.36 V PGA combining single-stage OTA and input negative transconductor for low energy RF receivers. Electron. Lett. **54**(5), 319–320 (2018). https://doi.org/10.1049/el.2017.4464
13. S. Zeller, C. Muenker, R. Weigel, T. Ussmueller, S. Member, A 0.039 mm2 inverter-based 1.82 mW 68.6 dB-SNDR 10 MHz-BW CT-SD ADC in 65 nm CMOS using power- and area-efficient design techniques. IEEE J. Solid-State Circ. **49**(7), 1548–1560 (2014)

Chapter 4
Design Methodology for ULV Circuits

4.1 CAD Tools for Analog Circuit Design

The design of analog integrated circuits requires the execution of several steps to convert the general circuit idea or a functional definition in a physical circuit. The design steps are classically divided in system, circuit and layout levels [3]. In the system level, the design task is related to the definition of the block diagrams and the individual specifications of each functional block. In the circuit level, the circuits topologies are chosen from available options that can satisfy the functional block specification. Based on that, each one of the bias voltage and bias current levels are chosen and the circuit devices are sized. The sizing phase is one of the most complex and hardworking task of an integrated circuit design because the analog designer should deal with the device modeling [24], several specification trade-offs [4] and to find solutions that are robust to the PVT variations [11]. In the layout level, the design is performed through the representation of the device physical layers and the interconnection between each circuit and to I/O PADs. At this level some layout techniques should be considered to reduce the mismatch and process variation effects [6, 11], to reduce the values of parasitic resistance and capacitance and other effects present in deep-submicron and nanometer technologies, such as the length of diffusion (LOD) and the well proximity (WPE) effects [16].

Several analog design tools have been proposed in the literature from the eighties to now [2, 7, 15, 21, 23, 25] but the analog design are still predominately performed using manual approach, some CAD tools to the schematic and layout draw, electrical simulators, design rule checks and post-layout parasitic extractions. The circuit sizing step is, in general, performed first by a hand simplified equation analysis [1, 18] or some bias-based look-up tables [13] to obtain the preliminary device sizes. The device sizes are refined to reach the target behavior on the circuit by performing several iterations of size adjustments and electrical simulations.

© The Author(s), under exclusive license to Springer Nature Switzerland AG 2022
L. C. Severo, W. A. M. Van Noije, *Ultra-low Voltage Low Power Active-RC Filters and Amplifiers for Low Energy RF Receivers*,
https://doi.org/10.1007/978-3-030-90103-5_4

4.2 Transistor Sizing of ULV Circuits

The ULV circuits proposed in this work are implemented using only two-stacked transistors to improve the output swing voltage and to operate with reduced supply voltages. Hence, both PMOS and NMOS transistors have the source terminals connected to a constant and well-known DC voltage (V_{DD} or $ground$). This characteristic reduces the design complexity since the classical MOS modeling has the terminal voltages referred to the source terminal [24]. Figure 4.1a shows the schematic of a single CMOS inverter circuit, used as the basic building block of all the circuits proposed in this work. It has the input connected to both transistors gate terminals, the output connected to both drain terminals and the bulk terminals are forward biased by the Vbp and Vbn voltages. The input and output DC voltages are defined by the common-mode voltage employed in the circuits. In the proposed applications, these voltages are assumed to be equal to $V_{DD}/2$ to maximize the output voltage swing, to avoid the DC currents flow when in closed-loop and to present similar overdrive voltages in both the NMOS and PMOS transistors. Thus, the voltage at the gate and drain terminals of the proposed circuits are equal to $V_{DD}/2$. The bulk DC voltages are also tied to $V_{DD}/2$ when it is adjusted by some feedback or replica circuit to have the maximum controllability margin from 0V to V_{DD}. When the bulk terminals are not adjusted, they are tied to V_{DD} or $ground$, according to the transistor type and the threshold voltage needed.

Based on the DC voltage levels analysis we can conclude that the proposed two-stacked transistors ULV circuits have well defined and constant bias voltages for all the transistors terminals related to the power supply voltage level. Once the V_{DD} voltage is defined, only the transistors channel width (W) and length (L) can be designed to reach the circuit target specifications values. Figure 4.1b shows the CMOS inverter representation by using individual gate to source (V_{GS}), drain to source (V_{DS}) and bulk to source (V_{BS}) voltage sources. As V_{GS}, V_{DS} and V_{BS} are defined by the $V_{DD}/2$ voltage level, each transistor can be individually sized, considering the same drain current (I_D) for both transistors. In other words, if a target drain current ($I_{D_{ref}}$) is defined to the circuit, the W/L aspect ratio of each transistor can be obtained. Moreover, all the specifications related to the current level, such as the small-signal transconductances (g_m and g_{mb}) and conductance (g_{ds}) can also be defined as reference values to obtain the W/L aspect ratio. The best reference design parameter varies from circuit to circuit and can be mixed during the circuit design. For example, the simplified negative transconductance shown in Fig. 3.7 can be designed using both $I_{D_{ref}}$ and $g_{m_{ref}}$ references. Transistors M1a/b should be designed to present the target negative transconductance ($g_{meq_{dm}}$) given by Eqs. (3.24) and (3.25). Using the g_m/I_D ratio of M1a/b, the current needed by transistors M2a, M2b and M2c can be found. Then, these transistors are designed to present the target drain current needed by M1a/b. The design procedure of the negative transconductor circuit will be detailed in Sect. 4.3.1.

The transistor channel length (L) is one of the most important design parameters in sub-micron and nanometer technologies, as previously analyzed in Chap. 2.

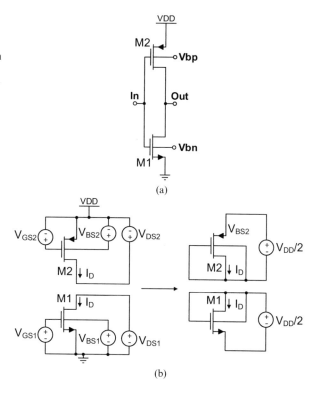

Fig. 4.1 Inverter-based circuit design: (**a**) schematic of the basic cell and (**b**) the individual bias representation

It has influences on the transistor threshold voltage, small channel effects, noise contribution and on the transistor process and mismatch variabilities. In addition to that, without the prior L definition, the transistor characteristics cannot be wholly defined to allow the W/L calculation. Thus, the L of each transistor should be used together with the reference drain current or transconductance to obtain the actual W/L ratio of ULV circuits.

The W/L aspect ratio of some transistor in ULV circuits must be high because of the low current density ($I_D/(W/L)$) when operating at the weak or moderated channel inversion level, as shown in Sect. 2.1.1. In such case, the use of parallel associated transistors is required to improve the layout regularity and to reduce the polysilicon gate resistance. Additionally, the parallel transistor match is necessary to implement the current mirrors used on the circuits to bias the current sources or control the node voltage level. Figure 4.2 shows a layout example of CMOS transistors design using multifinger and parallel associated devices. The shallow trench isolation (STI), employed in such modern technologies, changes the transistor parameters due to the mechanical stress in the diffusion region that increases with the length of the diffusion region. As a consequence, the multifinger devices have higher diffusion length (DL) and does not have the same behavior of a parallel associated single device [16]. To overcome this problem the use of multiple parallel associated devices is preferred to the multifinger design. This comfiguration also

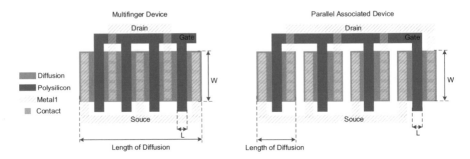

Fig. 4.2 Multifinger and parallel associated devices layouts

improves the layout regularity and make easier the design of interdigitated and common-centroid layouts. The main disadvantages of using multipliers instead of multifinger devices are the increased sidewall parasitic capacitance and the silicon area.

4.3   Proposed Operation-Point Simulation-Based Design Tool

The ULV circuits proposed in this work have well defined DC voltages related to the used V_{DD}, as shown in last section. The transistor sizing can be performed by using some explicit math expression that models the transistor drain current or the small-signal transconductance related to the device sizes.

However, the CMOS transistor modeling is not a simple task in sub-micron and nanometer CMOS processes due to several non-linear effects, related to the fabrication process complexity. Additionally, at the ULV range, the transistors are operating in the moderate or weak channel inversion levels in which both drift and diffusion charge carriers should be considered on the device current conduction modeling. Because of that, the drain current expression is not only dependent on the biasing and transistor sizing, but also on several process dependent parameters (p_i), as shown in Eq. (4.1) [19, 24]. Furthermore, in modern CMOS processes, each one of the model parameters (p_i) is not a constant value but conventionally defined as a piecewise function related to the transistor sizes, as shown in Eq. (4.2). The foundries use this parameter extraction strategy to preserve the modeling reliability in all the possible W and L value combinations. As a consequence of the modeling complexity, it is not possible to design the circuit directly using a drain current expression that results in the W and L, since the model parameter values are dependent on the W and L or, in other words, no accurate explicit function exists as the f function of Eq. (4.1).

$$I_D = f(W, L, V_{GS}, V_{DS}, V_{BS}, p_1, p_2, \ldots, p_n) \qquad (4.1)$$

$$
p_i(W, L) = \begin{cases} value_{11} & W_1 \leq W < W_2 \ and \ L_1 \leq L < L_2 \\ value_{21} & W_2 \leq W < W_3 \ and \ L_1 \leq L < L_2 \\ \quad \vdots \\ value_{i1} & W_{i-1} \leq W < W_i \ and \ L_1 \leq L < L_2 \\ value_{12} & W_1 \leq W < W_2 \ and \ L_2 \leq L < L_3 \\ value_{22} & W_2 \leq W < W_3 \ and \ L_2 \leq L < L_3 \\ \quad \vdots \\ value_{ij} & W_{i-1} \leq W \leq W_i \ and \ L_{j-1} \leq L \leq L_j \end{cases} \tag{4.2}
$$

To address this issue and to contribute for the development of a generic design strategy for the ULV circuits sizing, a numerical-based design tool using commercial electrical simulators is also proposed in this work.

The proposed tool is structured by the implementation of Eq. (4.1) through an operation-point simulation using a SPICE electrical simulator. Thus, the design can be performed using the complete simulation models available on the process design kit (PDK) provided by the foundry, making the design task faster without the need for a complete device modeling and parameter set extraction. Additionally, the SPICE operation-point simulation provides other important bias dependent parameters, such as the small-signal transconductances and the parasitic capacitances, as illustrated in Eq. (4.3). These extra parameters can be used during the design phase to expand the applicability of the proposed tool.

$$
[I_D, g_m, g_{ds}, g_{mb}, \ldots, c_{gs}, c_{gd}, c_{ds}] = f_{OPsim}(W, L, V_{GS}, V_{DS}, V_{BS}, p_1, p_2, \ldots, p_n) \tag{4.3}
$$

The SPICE operation-point simulation uses the information of the transistor channel length (L) and width (W), voltage bias (V_{GS}, V_{DS} and V_{BS}) and the model parameters ($p_1, p_2, \ldots, p_n$). Therefore, it is only appropriate for a device behavior check whenever the transistor sizes and the bias voltages are known. Thus, such expressions as shown in Eqs. (4.4) and (4.5), are more appropriate for the ULV transistor sizing. In these expressions, the function f_{CalcW} receives the transistor L, bias voltages and the target reference drain current ($I_{D_{ref}}$) or reference transconductance ($g_{m_{ref}}$) and returns the respective transistor width and all the operation-point information ($Opinfo$). Using these equations all the transistor from the ULV circuit can be designed for a given L.

$$
[W, Opinfo] = f_{CalcW_{I_D}}(I_{D_{ref}}, L, V_{GS}, V_{DS}, V_{BS}, p_1, p_2, \ldots, p_n) \tag{4.4}
$$

$$
[W, Opinfo] = f_{CalcW_{gm}}(g_{m_{ref}}, L, V_{GS}, V_{DS}, V_{BS}, p_1, p_2, \ldots, p_n) \tag{4.5}
$$

The implementation of Eqs. (4.4) and (4.5) were performed by means of an iterative numerical function, using the operation-point simulation results. In other words, the function of Eq. (4.3) is analyzed some times to find the W parameter value that results in the target $I_{D_{ref}}$ or $g_{m_{ref}}$ value. Figure 4.3 shows the flowchart used to implement the $f_{CalcW_{I_D}}$ function of Eq. (4.4). The function receives as input parameters $I_{D_{ref}}$, L and the bias voltages (V_{GS}, V_{DS} and V_{BS}). The model parameters $p_1, p_2, \ldots, p_n$ are directly included to the SPICE simulation by using a model library file. The transistor model information, the path to the library file and all the tools settings (start point, tolerated error, grid, and other) are transmitted to the tool using a Configuration File. The flow starts by resetting the iteration counter (Count) and by using a start point channel width (W_0), defined to the minimum value allowed by the fabrication process or another intermediate value to reduce the number of iterations needed in the convergence. In the next step, the simulation netlist text file is written, including a single transistor (NMOS or PMOS), the bias voltage sources and some SPICE directives. After that, the electrical simulation is ran and the operation-point information (Op info) is saved in a text file. This file is read in the next step, and the simulated drain current (I_{DS}) is extracted. The I_{DS} value is compared to $I_{D_{ref}}$ and the maximum percent current error tolerated (e_{rr}). After that, the W value is updated by the factor $I_{D_{ref}}/I_{DS}$ and the iterative process is repeated while the maximum error or the maximum number of iteration (N) is not satisfied. The algorithm returns the calculated W and the operation-point information of the designed transistor at the end of the algorithm execution.

Some extra steps are performed by the proposed function implementation, not shown in Fig. 4.3, to adjust the W value to the fabrication process grid and the use of multipliers and multifingers when high W/L aspect ratio transistors are needed. Further, a step is added to the function to allow the design using series-parallel transistor association [9] that is a very important design strategy for low-frequency (Kilo-Hertz range) ULV circuits [5, 8] but it was not employed in this work due to the operation in higher frequencies (Mega-Hertz range). A variation of the algorithm depicted in Fig. 4.3 is used to implement Eq. (4.5), but using $g_{m_{ref}}$ and comparing it to the simulated transistor transconductance. An optional setting is also added to the function to allow the W calculation using a constant number of parallel transistors (M) or fingers (Nf). It can be used to obtain the value of W where a certain number of parallel devices should be considered for layout design or to improve the transistor matching. The use of this option will be detailed in Sect. 4.3.1.

The proposed algorithms were implemented on the Matlab® environmental as a toolbox of functions. These functions can be used to implement design scripts to size all the transistors of a ULV circuit. As the functions return all the operation-point information, this data can also be processed and considered to estimate the circuit specifications by utilizing some circuit modeling equations and hand simplified expressions, such as the amplifier voltage gain, bandwidth, and the input-referred noise specifications. The Synopsis HSpice® electrical simulator is used in the proposed tools. It was chosen because it is compatible with several PDKs, is widely used by the microelectronics designers and does not require a complete simulation environmental configuration. However, the tool implementation has a generic text-

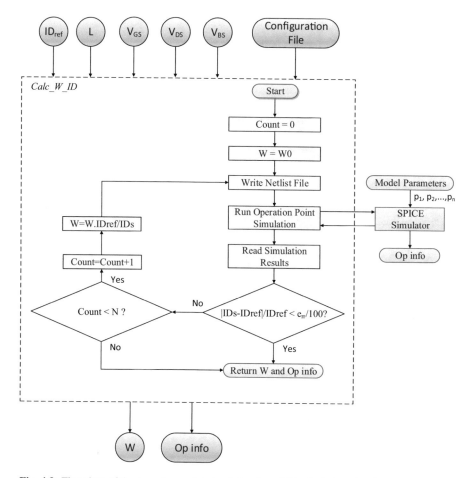

Fig. 4.3 Flowchart of the operation-point simulation-based W calculation algorithm using a drain current reference ($I_{D_{ref}}$)

based simulation interface that makes possible the use of other commercial electrical simulators, such as the Cadence Spectre® and Mentor Graphics Eldo® and freeware simulators, such as the Analog Devices LTspice® and NGSpice.

We have also implemented in the tool the graphical user interface (GUI) shown in Fig. 4.4 to make easier and simpler the use by students, designers and researchers. The GUI shows to the user all the operation-point information and other ULV essential design parameters, such as the g_m/I_D and g_{mb}/g_m ratios, the current densities and the actual threshold voltage. Additionally, a parameter sweep environmental GUI was added to the tool where it is possible to sweep a design parameter and to plot the sweep effect on the operation-point values. It is very useful to help the user in the definition of the design parameter values, as the best values for the transistor length (L) and the number of parallel associated devices. Figure 4.5a

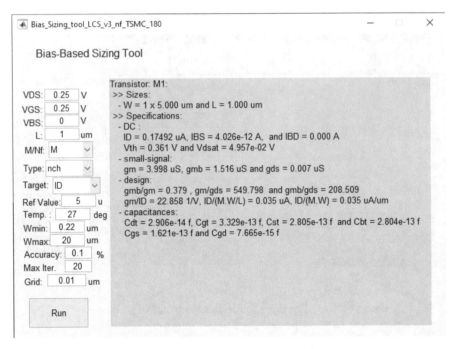

Fig. 4.4 Operation-point simulated-based sizing tool in a graphical user interface (GUI)

and b show the parameter sweep GUI environmental and a sample curve obtained for the threshold voltage (V_T) variation of the channel length sweep in a 180 nm CMOS process. In the sample curve, we can see the reverse short-channel effect (RSCE) of a standard-V_T device.

The proposed tool has some advantages in comparison to other tools from the literature. The tool uses the model parameters set in a commercial and wide used electrical simulator and is compatible with all the fabrication process in which a PDK has HSpice models. This is the main advantage in comparison to other tools proposed in [23] and [10] that are not related to the PDK parameters. The proposed tool does not need extra data or simulations to obtain some abacus or look-up tables, which is necessary in the [13] implementation. The disadvantage of the proposed tool is the need for a commercial electrical simulator. However, the HSpice simulator is one of the most commonly used electrical simulators and it is often available in design houses, research centers and universities.

4.3.1 Design Example

In this section a design methodology using the proposed tool to size the transistors of the ULV circuits is exemplified. The developed design methodology is used to

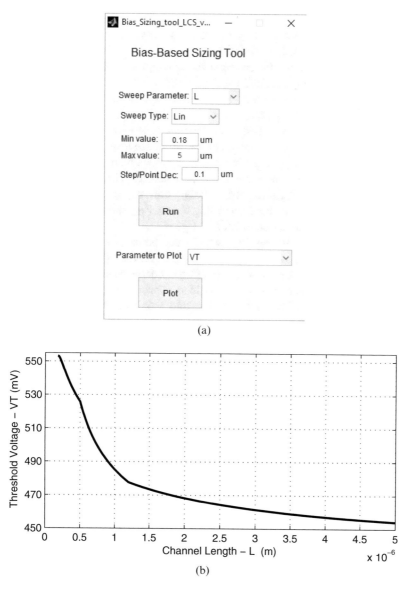

Fig. 4.5 Parameter sweep environmental in the implemented tool (**a**), and a sample of sweep variation curve obtained for the threshold voltage in function of the channel length variation (**b**)

design some of the programmable gain amplifiers and active filters for low energy RF receivers presented in Chap. 5.

The OTA design is exemplified by using the proposed circuit shown in Fig. 3.16 of Sect. 3.3.1. In this design, transistors M5a to M5d have the same W and L values equal to W_5 and L_5, respectively. The same is defined to the M6a to M6d transistors, using the W_6 and L_6 parameters. The difference among these transistors is the multiplicity defined as M_{ab} to M5a/b and M6a/b and M_{cd} to M5c/d and M6c/d. Thus, the design free variables are W_5, L_5, W_6, L_6, M_{ab} and M_{cd}. The circuit can be designed in different ways using the proposed operation-point based tool and the specification required for the amplifier. Here, it will be designed using the gain-bandwidth product (GBW) and the common-mode rejection ratio (CMRR) specification values as design references.

The first strategy is performed with the design of unitary transistors to obtain the W_5 and W_6 values, considering a current reference level (I_{D1}) and the value of L_5 and L_6. With the unitary transistors, Ma/b can be manually adjusted to satisfy the required GBW value while Mc/d can be changed to obtain the target CMRR. Figure 4.6 shows the single transistor design flow using the proposed tool. The transistor voltage bias is defined according to the V_{DD} used and the $Calc\_W\_I_D$ function is applied twice to obtain W_5 and W_6.

An improved design strategy for the OTA can be performed using an equation-based approach together with the proposed tool. The OTA GBW can be estimated by simplifying Eq. (3.33), as shown in Eq. (4.6). It depends on the transconductance of transistors M5a/b and M6a/b and the load capacitance (C_L). Using the target GBW and the C_L values, transistors M5a/b and M6a/b can be design to obtain the equivalent transconductance $g_{m5} + g_{m6}$ equal to $\omega_{GBW}.C_L$. The design of M5a/b and M6a/b is easily performed by using a reference drain current level ($I_{D_{GBW}}$) for both transistors instead of using g_{m5} and g_{m6}, since g_{m5} is not equal to g_{m6}. This current can be estimated from Eq. (4.6), using the g_m/I_D ratio of M5a/b and M6a/b, as shown in Eq. (4.7).

$$\omega_{GBW} = \frac{g_{m5} + g_{m6}}{C_{io} + C_o + C_{ob}} \approx \frac{g_{m5} + g_{m6}}{C_L} \tag{4.6}$$

Fig. 4.6 Flowchart of the design methodology used in the transistor sizing of the OTA shown in Fig. 3.16

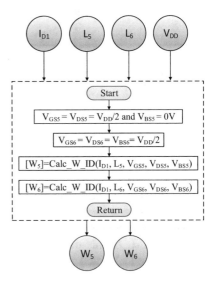

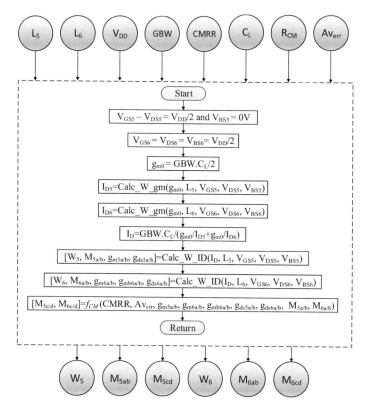

Fig. 4.7 Flowchart of the design methodology used in the transistor sizing of the OTA shown in Fig. 3.16—improved version

$$I_{D_{GBW}} = \frac{\omega_{GBW} . C_L}{\left(\frac{g_m}{I_D}\right)_5 + \left(\frac{g_m}{I_D}\right)_6} \tag{4.7}$$

Figure 4.7 shows the flowchart of the OTA improved design methodology. In this flow, the transistor bias voltages are calculated using the V_{DD} voltage. A first approximation using $g_{m5} = g_{m6} = g_{m0}$ in Eq. (4.6) is used to obtain a start point transconductance $g_{m0} = \omega_{GBW} . C_L / 2$. This value is used in function $Calc\_W\_g_m$ to design M5a/b and M6a/b and to obtain the drain current needed in each transistor to present a transconductance equal to g_{m0}. The current information is employed to calculate the g_m/I_D ratio of these transistors. Thus, Eq. (4.7) is used to obtain the needed drain current of M5a/b and M6a/b and the function $Calc\_W\_I_D$ is applied twice to obtain the W and the multiplicity of transistors M5a/b and M6a/b. The information of the transconductance and output conductances of M5a/b and M6a/b are captured from $Calc\_W\_I_D$ to obtain the multiplicity of M5c/d and M6c/d needed to satisfy the target CMRR. Combining Eq. (3.43) with Eq. (3.44), Eq. (4.8) is obtained to estimate the CMRR. The values of g_{mcm} and g_{dseq} are defined by Eqs. (3.41) and (3.42), which are rewritten by Eqs. (4.9) and (4.10), considering the multiplier M_{ab} and M_{cd}. Using these equations, Eq. (4.11) is obtained for the M_{cd} calculation using M_{ab} and the operation-point information of transconductances and output conductances. Thus, whole the OTA design free variables are designed to attend the target GBW and CMRR.

$$CMRR = \frac{Av_{dm0}}{Av_{cm0}} = \frac{g_{dseq} + Av_{err} \cdot g_{mcm}}{g_{dseq} + 1/R_{cm}} \approx \frac{Av_{err} \cdot g_{mcm}}{g_{dseq} + 1/R_{cm}} \tag{4.8}$$

$$g_{mcm} = g_{mb6_{ab}} + g_{mb6_{cd}} + g_{m5_{cd}} + g_{m6_{cd}} = g_{mb6_{ab}} + \frac{M_{cd}}{M_{ab}} \cdot \left(g_{mb6_{ab}} + g_{m5_{ab}} + g_{m6_{ab}} \right) \tag{4.9}$$

$$g_{dseq} = g_{ds5_{ab}} + g_{ds6_{ab}} + g_{ds5_{cd}} + g_{ds6_{ab}} = g_{ds5_{ab}} + g_{ds6_{ab}} + \frac{M_{cd}}{M_{ab}} \cdot \left(g_{ds5_{ab}} + g_{ds6_{ab}} \right) \tag{4.10}$$

$$M_{cd} \approx M_{ab} \cdot \frac{CMRR. \left(g_{ds5_{ab}} + g_{ds6_{ab}} + 1/Rcm \right) - Av_{err} \cdot g_{mb6_{ab}}}{g_{mb6_{ab}} + g_{m5_{ab}} + g_{m6_{ab}} - g_{ds5_{ab}} - g_{ds6_{ab}}} \tag{4.11}$$

The negative transconductor design exemplification is performed by using the proposed circuit shown in Fig. 3.9, of Sect. 3.2. In this circuit, all the transistor and the Rex, external resistor, should be sized to obtain the target equivalent negative transconductance (g_{mneg}). The same current level is considered for the main transconductor, the replica bias and the constant g_m bias. Therefore, the following transistor equality are assumed: M1a=M1b=M1r=M1x and M2a=M2b=M2c=M2d=M2r=M2=M2x.[1] Transistor M7a and M7b have the same current level, but due to the voltage drop on Rex (of ΔV_{Rex}) they are designed to have the same W and L but using different multiplicities (M_{7a} and M_{7b}). Based on that, the design of the negative transconductor has the following design variables: W_1, L_1, M_1, W_2, L_2, M_2, W_7, L_7, M_{7a}, M_{7b} and Rex. Where W_1, L_1 and M_1 are the parameters of M1x and W_2, L_2 and M_2 are the parameters of M2x.

Figure 4.8 shows the design methodology using the developed tool to design the negative transconductor. Based on the V_{DD} voltage, the bias voltages of M1x and M2x are calculated. As a start point, the simplification $g_{m1} = -g_{mneg}$ is assumed to obtain $W1$ and M_1 using the $Calc\_W\_gm$ function and the g_{m1} value. Transistor M2x is designed to adopt the I_{ds1} current obtained from the operation point information of the M1x design. After the design of M1x and M2x, the simulated negative transconductance (g_{neg_s}) is evaluated using Eq. (3.24) and the operation point transconductance of M1x and the output conductances of M1x and M2x. After that, g_{neg_s} is compared with the target negative transconductance, g_{m1} is updated to the desired value and the loop is executed again while the calculated error is higher

---

[1] M1x and M2x are referred only to give generic names for the transistors and are not physical devices.

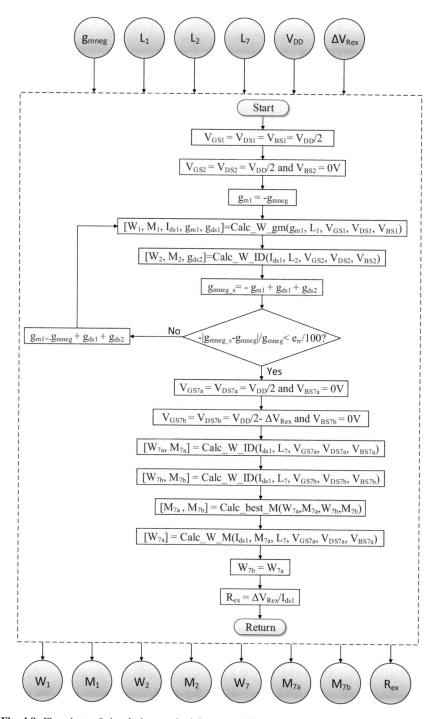

Fig. 4.8 Flowchart of the design methodology used for the transistor sizing of the negative transconductor shown in Fig. 3.9

than the tolerated error (e_{rr}). At the end of this loop, transistors M1x and M2x are completely defined and the drain current reference (I_{ds1}) is known. In the next step, the M7a and M7b PMOS transistors of the constant g_m bias circuit are designed. The bias voltages of M7a and M7b are obtained using the V_{DD} and ΔV_{Rex} values. Both transistors sizes are calculated using the drain current reference I_{ds1} and, as M7a and M7b have different voltage bias, the W and M obtained are different from each other. With these values, function $Calc\_best_M$ is employed to define the best ratio for the multiplicity factor of M7a and M7b that best approximate the obtained W_{7b}/W_{7a} ratio. At this point, for the multiplicity calculation, function $Calc\_W\_M$ is used to obtain the W_{7a} considering the reference drain current and the fixed best multiplicity ratio. The obtained W is adopted for both M7a and M7b transistors. The external resistor Rex is sized using the ohm's law with the values of the ΔV_{Rex} and the branch reference current (I_{ds1}). At the end of the flow execution, all values for the design variables are obtained and the proposed negative transconductor circuit is wholly sized.

The error amplifiers used in the CMFB and replica bulk bias can be designed using the same strategy employed on the OTA design. The CMFB error amplifier should be designed to present a reasonable bandwidth and reduced input parasitic capacitance. On the other hand, the replica bias error amplifier should be designed to present a reduced power dissipation and a low sensitivity to the mismatch variability.

4.4   ULV Circuit Design Using the UCAF Tool

The design methodology based on the operation-point simulation, presented in the last section, is very powerful to design ULV circuits. However, a prior definition is required to the transistor channel length (L) and the current level, or the small-signal transconductances. Thus, the designer should analyze the device behavior and the circuit specification equations before the circuit design to find the best values of these parameters.

To improve the circuit design with no need of predefined parameters we have also used in this work an improved version of the UCAF tool. UCAF is an optimization-based tool developed at the Computer Architecture and Microelectronics Group (GAMA) of the Federal University of Pampa [21]. It is an analog integrated circuit sizing tool that includes some functional blocks that can be configured to design any kind of analog circuit, as shown in Fig. 4.9. Additionally, it includes some special design strategies to find solutions with low sensitivity to the process, voltage and temperature (PVT) variations using optimized Monte Carlo [22] and process corner [20] simulations.

The CMOS transistor channel width (W) and length (L) are the standard analog circuit design variables of the UCAF tool. The simplified design space exploration flow is illustrated in Fig. 4.10. The optimizer generates the values for each one of the design variables (W and L) and according to the solution quality it explores the design space to obtain high-quality solutions (or optimized solutions). The UCAF

Fig. 4.9 UCAF modular functions. Source: Adapted from [21]

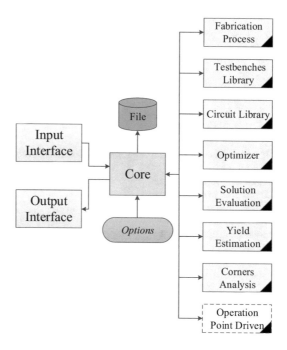

tool includes the Genetic Algorithms, Simulated Annealing and Particle Swarm as global optimization meta-heuristics and the Sequential Quadratic Programming and the Nelder–Mead as local optimization methods. The local and global optimization methods can be configured to work individually or in a hybrid design exploration strategy. The solution quality analysis is obtained with the Solution Evaluation function. This function receives the values of the W and L variables and returns a cost function (f_c) value to the optimizer. In the UCAF tool, the cost function implementation is based on the use of a multi-objective to mono-objective weighted sum function [20]. The cost function calculation is given by the comparison of the circuit specifications values of the generated solutions with the target values for each specification. The circuit specifications of the generated solutions are estimated by using electrical simulations and some standard circuit testbenches. The UCAF tool uses the Synopsis HSpice® as the standard electrical simulator.

The use of only W and L parameters as design variables is essential in general purpose tools, such as the UCAF tool, because no more information is needed from the circuit under design and it can be seen by the design tool as a black box. Due to its flexibility, the use of only transistor sizes as design variables is also widely used on other tools presented in the literature [15, 17, 25].

However, the use of W and L as design variables is not efficient for ULV designs. As shown in Sect. 4.2, the ULV circuits have well-defined voltage bias that are dependent on the V_{DD} power supply voltage. Because of that, only a few W and L combination results in the appropriated DC voltage bias and are practical. Thus, a high number of unfeasible solutions are generated during the optimizer design

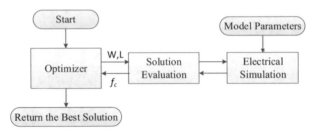

Fig. 4.10 The simplified design space exploration flow of the UCAF tool

Fig. 4.11 Design variable
comparison in a single-ended
CMOS inverter amplifier: (**a**)
conventional tool and (**b**)
proposed operation-point
driven tool

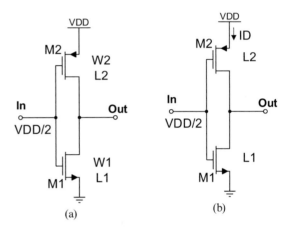

space exploration. This is illustrated in Fig. 4.11a for a single-ended ULV CMOS
inverter amplifier. The conventional design variables are the W and L of transistor
M1 and M2, resulting in the four design free variables $W1$, $L1$, $W2$ and $L2$. The
input DC voltage is defined to the optimal common-mode voltage equal to $V_{DD}/2$.
The W and L parameters values for transistors M1 and M2 can be chosen to be
between the minimum and maximum bounds of the fabrication process, but only
a few combinations of them make the circuit practical. For example, if the aspect
ratio (W/L) of transistor M2 is chosen by the optimizer to be much higher than the
W/L ratio of M1, M1 will work in the saturation region while M2 will operate in
the linear region, degrading the amplifier performances and the output DC voltage
will tend to V_{DD}. This solution is not feasible and can not be used to an amplifier
circuit. Sometimes, this kind of solutions cannot even be evaluated by the simulation
testbenches, resulting in fail solutions. To be considered as a feasible solution,
transistor M1 and M2 should have the proper aspect ratio to present similar drain
current conduction and make the output voltage near to the $V_{DD}/2$ level.

The design exploration efficiency can be improved by including some information from the designed circuit to avoid some of the unpractical solutions. Some
strategies based on the operating-point analysis, known as operation-point driven
(OPD), were reported in the Literature by [14] and by [12]. It uses the bias point
information during the optimization procedure to reduce the number of unpractical

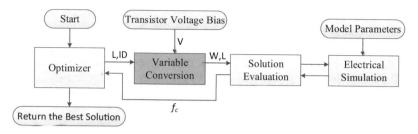

Fig. 4.12 The improved UCAF flow using the proposed operation-point driven strategy

solutions and also to reduce the number of design variables, making the optimizer exploration more efficient. Such kind of information can be easily inserted in the ULV circuit design exploration phase since the bias voltages are well known and the approximations needed on general circuits, as presented by [12], are not needed.

In this work, we propose the use of a mix of transistor sizes and OPD to make the design of ULV circuit more efficient. It is based on using the channel length (L) and the drain current (I_D) as the optimizer variables. Due to the series association present in all the two-stacked transistors circuits, the PMOS and NMOS transistors have the same drain current. Thus, it is possible to reduce the number of variables in the optimizer. This characteristic is significant in optimization-based tools because the design space dimension can be decreased, reducing the computation effort needed for the exploration. An example is illustrated in Fig. 4.11b for the single-ended CMOS inverter amplifier, in which the drain current and the channel length are used as design variable instead of the W and L. For this reason, the number of design variables is reduced from four ($W1, L1, W2, L2$) to three ($L1, L2$ and I_D).

The UCAF tool uses electrical simulations to evaluate each one of the optimized generated solutions. Thus, to implement the proposed mix design strategy, a variable conversion step is needed between the optimizer and the solution evaluation stages to provide the transistor size information to the electrical simulation. Figure 4.12 shows the design flow of the improved UCAF tool. The L and I_D optimizer variables values should be converted to W and L values before the solution evaluation. The variable conversion function can be performed using the proposed operation-point simulation-based tool, presented in Sect. 4.3. As the drain current of a CMOS transistor is directly related to the transistor aspect ratio, the operation point analysis can be performed to obtain the W from the L and I_D values, considering that all the transistor bias voltages are known. This step can be executed to design each one of the transistors individually, using the bias voltages and the I_D and L defined by the optimizer. As a result, all the optimizer variable values are converted to practical transistor sizes that have the needed bias voltage and the reference I_D current. This strategy is implemented, and an improved version of the UCAF tools was developed by inserting the Operation-Point Driven function, as shown in the dashed white box of Fig. 4.9.

The improved version of the UCAF tool is used to design some of the filters and amplifiers shown in Chap. 5. For the OTA and error amplifier designs, the L of each transistor and the current of each branch (with no matched currents) are used as design variables. In this case, the design flow of the variable conversion step is similar to the flow presented in Fig. 4.6 of Sect. 4.3.1, but the transistor multiplier calculation is performed using the required drain current level. The negative transconductor is designed using the same design flow shown in Fig. 4.8, but the UCAF tool is used to explore only the transistor L parameters. As the transistor length is related to several transistor characteristics, this exploration is very important in the negative transconductor design since it should be designed to present a reduced output noise and to present a reduced sensitivity to the process and temperature variations.

4.5 Conclusion

The two-stacked transistors ULV circuits proposed in this work have the bias voltages related to the V_{DD} voltage used. Thus, the operation-point simulation-based approach can be extensively used to design all the transistors of the ULV circuit.

A simple and useful tool, based on the operation-point simulation, is proposed to find the transistor channel width (W), considering the transistor channel length (L) and the drain current or the small-signal transconductance reference values. This tool can be applied to design all the circuits proposed in this work.

An improved version of the UCAF optimization-based tool is also proposed using the implemented operation-point simulation-based tool. It can be used to explore some of the design parameters to obtain optimized solutions for the circuits. The improvement increased tool efficiency on the ULV circuit design space exploration, because it reduces the number of design variables and also the number of unpractical solutions.

The design methodologies and tools presented in this chapter are applied to design the filters and programmable amplifier presented in Chap. 5. As the design tool is not the main objective of this work, the analysis presented in this chapter were simplified with the focus on its use, functionalities and applicability. However, it has been a very powerful tool to get rapidly a new solution for any of the designed circuits in this book.

References

1. P.E. Allen, D.R. Holberg, *CMOS Analog Circuit Design*, 2nd edn. (Oxford, New York, 2002)
2. K. Antreich, J. Eckmueller, H. Graeb, M. Pronath, F. Schenkel, R. Schwencker, S. Zizala, WiCkeD: analog circuit synthesis incorporating mismatch, in *Proceedings of the IEEE 2000 Custom Integrated Circuits Conference*, Orlando (IEEE, New York, 2000), pp. 511–514. https://doi.org/10.1109/CICC.2000.852720

3. S. Balkir, G. Dundar, A.S. Ogrenci, *Analog VLSI Design Automation*, vol. 650 (CRC press, Boca Raton, 2003)
4. D.M. Binkley, *Tradeoffs and Optimization in Analog Cmos Design*, vol. 2 (Wiley, Chichester, 2007)
5. R.A.S. Braga, H.C. Ferreira, G.D. Colletta, O.O. Dutra, Calibration-less nauta OTA operating at 0.25-V power supply in a 130-nm digital CMOS process, in *2017 IEEE 8th Latin American Symposium on Circuits & Systems (LASCAS 2015)* (2017), pp. 12–15
6. P.G. Drennan, C.C. McAndrew, Understanding MOSFET mismatch for analog design. IEEE J. Solid-State Circ. **38**(3), 450–456 (2003)
7. F. El-Turky, E. Perry, BLADES: an artificial intelligence approach to analog circuit design. IEEE Trans. Comput.-Aided Des. Integr. Circ. Syst. **8**(6), 680–692 (1989). https://doi.org/10.1109/43.31523
8. L.H.C. Ferreira, S. Member, S.R. Sonkusale, S. Member, A 60-dB gain OTA operating at 0.25-V power supply in 130-nm digital CMOS process. IEEE Trans. Circ. Syst. I: Regul. Pap. **61**(6), 1609–1617 (2014)
9. C. Galup-Montoro, M.C. Schneider, I.J.B. Loss, Series-parallel association of FET's for high gain and high frequency applications. IEEE J. Solid-State Circ. **29**(9), 1094–1101 (1994). https://doi.org/10.1109/4.309905
10. P. Giacomelli, M. Schneider, C. Galup-Montoro, MOSVIEW: a graphical tool for MOS analog design, in *Proceedings 2003 IEEE International Conference on Microelectronic Systems Education. MSE'03*, Anaheim (IEEE Computer Society, New York, 2003), pp. 43–44 https://doi.org/10.1109/MSE.2003.1205247
11. H.E. Graeb, *Analog Design Centering and Sizing* (Springer, Dordrecht, 2007)
12. I. Guerra-Gómez, T. McConaghy, E. Tlelo-Cuautle, Operating-point driven formulation for analog computer-aided design. Analog Integr. Circ. Signal Process. **74**(2), 345–353 (2013)
13. P.G.A. Jespers, B. Murmann, *Systematic Design of Analog CMOS Circuits* (Cambridge University Press, Cambridge, 2017)
14. B. Liu, M. Pak, X. Zheng, G. Gielen, A novel operating-point driven method for the sizing of analog IC. Proc. - IEEE Int. Symp. Circ. Syst. (1), 781–784 (2011). https://doi.org/10.1109/ISCAS.2011.5937682
15. N. Lourenço, R. Martins, A. Canelas, R. Póvoa, N. Horta, AIDA: layout-aware analog circuit-level sizing with in-loop layout generation. Integr. VLSI J. **55**, 1–14 (2016)
16. H.C. Ou, K.H. Tseng, J.Y. Liu, I.P. Wu, Y.W. Chang, Layout-dependent effects-aware analytical analog placement. IEEE Trans. Comput.-Aided Des. Integr. Circ. Syst. **35**(8), 1243–1254 (2016). https://doi.org/10.1109/TCAD.2015.2501293
17. R. Phelps, M. Krasnicki, R.A. Rutenbar, L.R. Carley, J.R. Hellums, Anaconda : simulation-based synthesis of analog circuits via stochastic pattern search. IEEE Trans. Comput.-Aid. Des. Integr. Circ. Syst. **19**(6), 703–717 (2000)
18. B. Razavi, *Design of Analog CMOS Integrated Circuits*, vol. 6 (McGraw-Hill, New York, 2001). https://doi.org/10.1111/j.1151-2916.1994.tb07040.x
19. M.C. Schneider, C. Galup-Montoro, *CMOS Analog Design Using All-Region MOSFET Modeling*, 1st edn. (Cambridge University Press, Cambridge, 2010)
20. L.C. Severo, W.A.M.V. Noije, An optimization-based design methodology with PVT analysis for ultra-low voltage analog ICs, in *Conference on PhD Research in Microelectronics and Electronics (PRIME)*, Lisbon (IEEE, New York, 2016), pp. 1–4
21. L.C. Severo, A. Girardi, A.B. de Oliveira, F.N. Kepler, M.C. Cera, Simulated annealing to improve analog integrated circuit design: trade-offs and implementation issues, in *Simulated Annealing - Single and Multiple Objective Problems* (Intech, London, 2012), pp. 261–283
22. L.C. Severo, F.N. Kepler, A.G. Girardi, Automatic synthesis of analog integrated circuits including efficient yield optimization, in *Computational Intelligence in Analog and Mixed-Signal (AMS) and Radio-Frequency (RF) Circuit Design* (Springer International Publishing, Cham, 2015), pp. 29–58
23. D. Stefanovic, M. Kayal, *Structured Analog CMOS Design*, vol. 53, 1st edn. (Springer Netherlands, Dordrecht, 2009)

24. Y. Tsividis, *Operation and Modeling of the MOS Transistor*, 2nd edn. (Oxford University Press, New York, 2003). ISBN:978-0-1951-7015-3
25. T.O. Weber, *Síntese de CIs analógicos em nível de circuito e sistema utilizando métodos modernos de otimização*. PhD thesis, Unversidade de São Paulo, 2015

Chapter 5
Design and Experimental Results

5.1 Complex Band-Pass Image-Rejection Filter

The complex band-pass filter (CxBPF) is a very important building block of modern Low-IF RF receivers. It is used to select the desired channel signal from the received signals and to reject the image signal generated after the down-conversion process. A CxBPF can be designed using two section of integrator-based low-pass filters, one for the in-phase (I) signal $S_I = |S|\angle 0°$ and other to the quadrature (Q) signal $S_I = |S|\angle +90°$. As the I and Q signals have a phase difference of $90°$, the transformation from low-pass to band-pass behavior is obtained using multiple feedbacks between the I and Q sections [12]. Thus, the low-pass filter real poles and the complex-conjugate poles centered in the real axis are moved to a complex position in the pole-zero diagram [9]. Additionally, due to the multiple-feedback between the I and Q signals, the desired signal (S_{sig}) is selected from the received signals while the image signal (S_{img}) is rejected. The ratio between S_{sig} and S_{img} in the pass-band is defined as the image-rejection ratio (IRR) of the CxBPF. The IRR of generic first and second order filters can be evaluated using Eq. 5.1 and Eq. 5.2, respectively [1]. By using these equations we can conclude that the higher the center frequency, the higher the IRR is. Additionally, it is possible to obtain higher IRR using a second order filter instead of using two cascaded first order filters due to the flexibility of the Q_{filter} choice in second-order filters.

$$IRR_{1st} = \sqrt{1 + 4.\omega_c^2/\omega_0^2} \tag{5.1}$$

$$IRR_{2nd} = \sqrt{\left(1 + 4.\omega_c^2/\omega_0^2\right)^2 + \left(4/Q_{filter}\right)^2 .\omega_c^2/\omega_0^2} \tag{5.2}$$

Where: ω_0 is the pole frequency, ω_c is the BPF center frequency and Q_{filter} is the conjugated-pole quality factor. ω_0 and Q_{filter} are defined by the LPF sections while ω_c is defined by the I/Q feedback.

The CxBPF designed in this work is based on the implementations presented in [3] and in [20], but using only a second order biquad and a first order filter to implement a third-order leapfrog active-RC filter. The schematic of the CxBFP designed in this work is shown in Fig. 5.1. It is composed of six single-stage OTAs divided into two sections of low-pass filters (LPF) to work with quadrature signals (I/Q). The low-pass filter sections control the bandwidth and the quality factor (Q_{filter}) according to the resistors (R) and capacitors (C1, C2 and C3) values. The bandwidth of the CxBPF is twice the cutoff frequency of each LPF section. The complex feedback provided by the resistors R_{IQ1}, R_{IQ2} and R_{IQ3}, changes the filter poles positions and generate the band-pass transfer function. Hence, the complex feedback resistors and the LPF capacitors control the center frequency (ω_c) that should be equal to the intermediate frequency (IF) of the BLE RF receiver. The CxBPF IF can be evaluated using Eq. 5.3 and it is used to design the complex feedback resistors, based on the designed LPF.

$$IF = \frac{1}{2.\pi.C_1.R_{IQ1}} = \frac{1}{2.\pi.C_2.R_{IQ2}} = \frac{1}{2.\pi.C_3.R_{IQ3}} \tag{5.3}$$

In the proposed CxBPF circuit, negative transconductors are placed at the input of each one of the single-stage OTAs in order to compensate the effect of the low voltage gain and the resistive load sensitivity, as previously analyzed in Sect. 3.1.

The following subsections present the filter passive devices design to find the proper CxBPF behavior and also the design of the negative transconductors and the OTAs. The circuit sizing was performed using the TSMC 180 nm design kit to operate with a power supply of 0.4 V by using low-V_T NMOS and PMOS transistors with 300 mV and 250 mV threshold voltage, respectively.

5.1.1 Filter Design

As presented in [19], the BLE 5 RF receiver should be designed to present a 1 MHz bandwidth, to reject the blockers interferences and to have a relaxed IRR of 24 dB in the 1 Mbps rate mode. The total receiver third-order intermodulation product (IIP_3) should be higher than -28 dBm to preserve the linearity requirements. The total receiver noise figure (NF) can be as high as 19 dB for a 15 dB SNR demodulator and, due to the LNA gain, the noise figure requirement for the CxBPF circuit is very relaxed. A third order BPF is sufficient to satisfy the rejection requirement of 32 dB at the adjacent channels in the 1 Mbps mode, but it is not sufficient to satisfy the rejection requirement of 41 dB in the 128 kbps mode. However, the design should consider the rest of the rejections in the receiver front-end. As presented in [13] and [16] after the down-conversion mixer a transimpedance amplifier (TIA) is used.

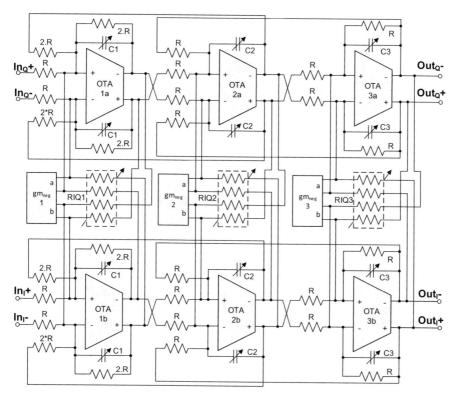

Fig. 5.1 Schematic of the designed third-order CxBPF. Reprinted, with permission, from [6]

Thus, the TIA add the first out of band attenuation and it is possible to satisfy the standard rejection requirement, for all the data rates, using a third-order filter at the baseband stage.

The filter center frequency—receiver IF—is defined in order to satisfy the standard requirement. For a bandwidth of 1 MHz, the minimum value of IF is 500 kHz. At this frequency, the circuit is optimized in terms of power dissipation, but it will suffer from a high flicker noise contribution at lower frequencies, a DC offset and a poor IRR. On the other side, a higher IF frequency improves the IRR and avoid the problems of DC offset and flicker noise, but it increases the filter power dissipation. To have a good compromise among IRR, power dissipation, DC-offset and flicker noise contribution an IF of two times the bandwidth is suggested by [9]. To further increase the IRR, the minimum bandwidth of 1 MHz is used to implement the filter. Thus, the BPF sections should have a cutoff frequency of 500 kHz and, consequently, the filter resistor R value, used as reference for all the resistors, and the capacitors C1, C2 and C2 can be obtained using the desired filter quality factor (Q_{filter}). Based on IF the values of the complex feedback resistors can be obtained according to Eq. 5.3.

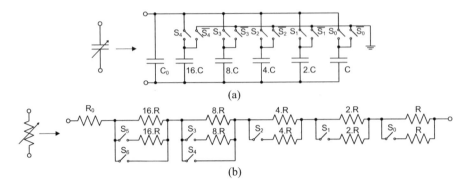

Fig. 5.2 Programmable capacitor (**a**) and resistor (**b**) used to implement the CxBPF

The complex feedback resistors and the capacitors were designed as programmable devices to provide a calibration capability on the bandwidth and IF. The schematic of the programmable capacitors and resistors are shown in Fig. 5.2a,b. The programmable capacitors were designed with a parallel association of five capacitor for C1 and six capacitors for C2 and C3. The capacitor value is changed using a digital signal that makes the switches on or off. The switches connect each capacitor to the parallel association or to the ground terminal in order to avoid floating internal nodes. The value of C was defined as 50 fF for all the capacitors, while C_0 was chosen according to the filter design and are different for each one of the capacitors C1, C2 and C3. The switches used for the capacitor association were implemented using low-V_T NMOS transistor with channel length and width of 0.3 μm and 8 μm and multiplicity of 8. The L is higher than the minimum value of 180 nm in order to reduce the RSCE on the threshold voltage.

The programmable resistors were implemented according to Fig. 5.2b using a hybrid series-parallel association. Usually, this kind of programmable resistors is implemented with short-circuits that trim-out some resistor from the series association. However, the switches at the ultra-low voltage operation, even with low-V_T devices, present a higher on-mode resistance (R_{on}) and the resistor short-circuit is not possible with small resistors values. Thus, we have used the switches to perform the parallel association in the smaller resistors (R, 2.R and 4.R) and the short-circuit switch is employed only in the higher resistors (8.R and 16.R). The switches used to program the equivalent resistor value were implemented using low-V_T devices with channel length and width of 0.5 μm and 5 μm and multiplicity of 10.

All the resistors and capacitors were sized to obtain the CxBPF target bandwidth and IF specifications. The programmability range of the capacitors and the complex feedback resistors were defined by the worst case scenario of process variability. The quality factor of the biquad filter was chosen to present the behavior of a third-order Butterworth BPF. Table 5.1 shows the values of each one of the passive devices used in the CxBPF implementation.

Table 5.1 Values of the passive devices used in the CxBPF implementation

Parameters	Typical value	Programmable range	Number of bits
R	100 kΩ	–	–
$C1$	1.1 pF	0.3 to 1.9 pF	5
$C2$	4.8 pF	1.6 to 8.0 pF	6
$C3$	2.4 pF	0.8 to 4.0 pF	6
R_{IQ1}	72.9 kΩ	45.4 to 100.0 kΩ	7
R_{IQ2}	16.7 kΩ	9.1 to 24.9 kΩ	5
R_{IQ3}	33.0 kΩ	21.2 to 49.3 kΩ	7

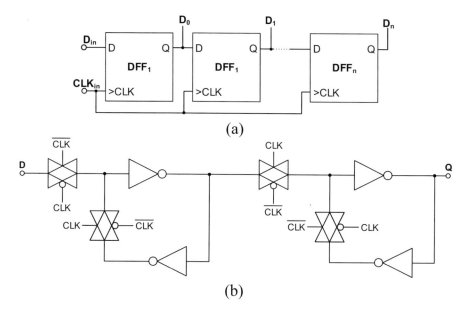

Fig. 5.3 Shift register used as serial input register bank (**a**) and the schematic of the edge-triggered D flip-flop (**b**) used in the implementation

In order to reduce the number of digital I/O pins needed to program the capacitors and resistors, we have designed a shift register that works as a series to parallel converter to configure all the filter digital bits. The shift register receives a digital serial input data that is converted to a parallel data after some clock cycles. The register bank was implemented using D type flip-flops, as shown in the schematic of Fig. 5.3a. The edge-triggered D flip-flop (DFF) was designed as shown in Fig. 5.3b, using CMOS transmission gates and inverters [2]. The transmission gates and inverters were designed using NMOS and PMOS standard devices with a channel length of 0.18 μm and width of 1 μm and 4.12 μm, respectively.

5.1.2   Negative Transconductors Implementation

The CxBPF implementation has an input negative transconductor for each one of the single-stage OTAs. The negative tranconductors were applied to compensate the OTA reduced voltage gain and the resistive load. The value of each one of the negative transconductance is dependent on the CxBPF resistors values. As previously analyzed in Sect. 3.1, the negative transconductance optimal value is equal to the inverse of the equivalent resistor obtained with the parallel association of all resistor connected to the input nodes.

Based on the CxBPF schematic of the Fig. 5.1 the optimal value for the g_{mneg_i} negative transconductance can be evaluated with Eq. 5.4, where the index i defines the filter stage and can be from 1 to 3. As the second stage of the CxBPF is an active integrator, a percentage safe margin of $\Delta g_{m\%}$ is added to the value obtained with Eq. 5.4 to avoid the instability risk. Based on the resistors typical values presented in Table 5.1 and using a safety margin of 10% for the second stage, the negative transconductances g_{mneg_1}, g_{mneg_2} and g_{mneg_3} should be equal to $-33.7\,\mu\text{S}$, $-71.9\,\mu\text{S}$ and $-50.3\,\mu\text{S}$.

$$g_{mneg_i} = -\left(\frac{2}{R} + \frac{1}{R_{IQi}}\right) \tag{5.4}$$

The negative transconductors used in the CxBPF implementation have the same topology presented in Sect. 3.2 and shown in Fig. 3.9a. The CxBPF I and Q sections should be as identical as possible because the mismatch between them reduces the IRR and changes the filter behavior. Thus, we designed the circuit shown in Fig. 5.4 that uses the same constant g_m bias and replica bias to implement both the negative transconductors used at the I and Q sections to reduce the mismatch effects and also to save power. The negative transconductor composed of M1a/b and M2a/b is connected to the input of OTAa, at the Q section, whereas the circuit composed of M1c/d and M2c/d is connected to the input of OTAb at the I section, as shown in Fig. 5.1. In this case, M1a/b=M1c/d and M2a/b=M2c/d.

Based on the needed values for each negative input transconductance, we have designed the negative transconductors to operate with V_{DD} of 0.4 V, using low-VT devices. A special attention was given to the negative transconductors sizing to reduce the noise contribution at the OTA inputs. For the sake of simplicity and to improve the layout regularity all the transistors channel length (L) were defined to be equal to 1 μm. This value was chosen in order to reduce the effect of the channel Halo implantation in the threshold voltage, as shown in Sect. 2.1.5, to reduce the transistor mismatch and to minimize the noise contribution, preserving the circuit area. Table 5.2 shows all the parameters used in the negative transconductor implementation. The error amplifier has the same transistor size for all the three negative transconductors, and it was designed to have a reduced bandwidth in order to keep the replica bias loop stable. Figure 5.5a–c show the layout of each one of the designed negative input transconductors.

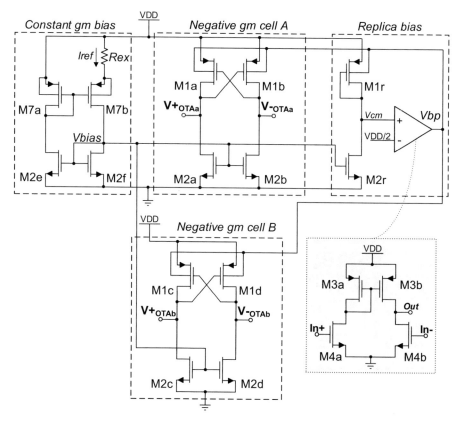

Fig. 5.4 Negative transcondutor used in the CxBPF implementation. Reprinted, with permission, from [6]

Table 5.2 Values of the parameters used in the CxBPF negative transconductors implementation

gm_{neg}	W (g_{mneg_1})	W (g_{mneg_2})	W (g_{mneg_3})	Unit
$M1a$ to $M1d$	3.78×1	9.42×1	5.69×1	μm
$M1r$	3.78×1	9.42×1	5.69×1	μm
$M2a$ to $M2f$	9.34×2	11.78×4	6.97×4	μm
$M2r$	9.34×2	11.78×4	6.97×4	μm
$M7a$	1.53×5	3.74×5	2.28×5	μm
$M7b$	1.53×14	3.74×14	2.28×14	μm
$M3a = M3b$	9.42×1	9.42×1	9.42×1	μm
$M4a = M4b$	3.76×1	3.76×1	3.76×1	μm
Other parameters	Value			Unit
$Rext_1$	25.19			kΩ
$Rext_2$	10.15			kΩ
$Rext_3$	16.75			kΩ

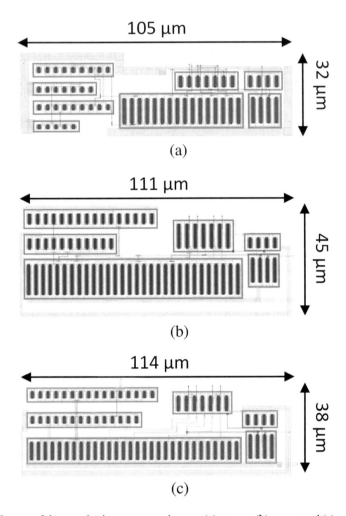

Fig. 5.5 Layout of the negative input transconductors: (**a**) g_{mneg_1}, (**b**) g_{mneg_2} and (**c**) g_{mneg_3}

5.1.3   OTA Implementation

The OTA used in the CxBPF implementation has the same topology presented in Sect. 3.3.1 and shown in Fig. 3.16. To reduce the design complexity, the same OTA implementation was used in all the six OTAs of the CxBPF.

The OTA unity gain frequency was designed to be over 14 MHz, in order to satisfy the $8 \cdot Q_{filter} \cdot f_{cutoff}$ relation, as suggested in [23], where Q_{filter} is the filter quality factor and f_{cutoff} is the highest cutoff frequency, equal to $1/\sqrt{2}$ and 2.5 MHz, respectively. Additionally to the unity gain frequency requirement, the design was performed in order to keep the common-mode and the power supply gains lower than 0 dB in all the pass band range.

The OTA was carefully sized using the improved version of the UCAF [18] to optimize the power dissipation and to be robust under process and mismatch variations. For the sake of simplicity, we have used all the transistor length equal to 1 μm and the design was performed using the transistor width (W) and the number of parallel transistors—multiplicity (M)—as design variables.

The common-mode sense resistors Rcm_a and Rcm_b were chosen in order to present a reasonable trade-off between the voltage gain reduction and to keep high the frequency of the pole generated with the input parasite capacitance of the error amplifier. The Cc capacitor was designed to make the CMFB phase margin higher than 45°. Table 5.3 shows all the transistors sizes and passive devices values used in the OTA implementation. The OTA layout is shown in Fig. 5.6 and it has the size of 85 μm×91 μm.

Table 5.3 Parameter values of all the transistor and passive devices used in the CxBPF OTA implementation

OTA parameters	W (OTA)	Unit
$M6a = M6b$	1.92×14	μm
$M6c = M6d$	1.92×8	μm
$M5a = M5b$	9.42×20	μm
$M5c = M5d$	9.42×2	μm
$M3a = M3b$	9.42×5	μm
$M4a = M4b$	3.76×5	μm
Other parameters	Value	Unit
$Rcma = Rcmb$	100	kΩ
Cc	0.8	pF

Fig. 5.6 Layout of the OTA used in the CxBPF implementation

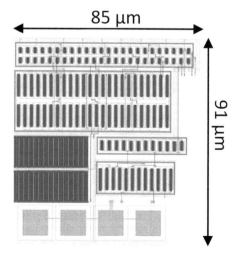

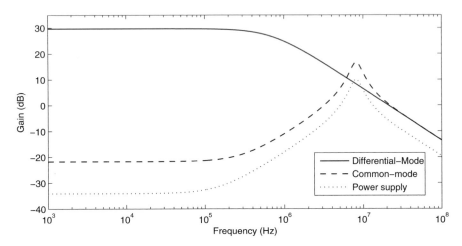

Fig. 5.7 OTA differential-mode, common-mode and power supply gains as function of the frequency. Reprinted, with permission, from [6]

Table 5.4 OTA post-layout simulation results

Specifications	Value	Unit
Technology	180	nm
Supply voltage	0.4	V
Differential-mode gain	29.66	dB
Unity gain frequency	21.88	MHz
Common-mode gain	−21.73	dB
CMRR	51.39	dB
Power supply gain	−34.08	dB
PSRR	63.74	dB
Slew Rate	9.92	V/μs
DC power dissipation	7.50	μW
Layout Area	0.0077	mm^2
Capacitive load	5	pF

The OTA layout parameters were extracted by the Cadence environmental, and post-layout simulations were performed, considering a capacitive load of 5 pF. Figure 5.7 shows the OTA differential-mode ($Av_{dm} = vo_{dm}/vi_{dm}$), common-mode ($Av_{cm} = vo_{cm}/vi_{cm}$) and power supply gains ($Av_{vdd} = vo_{cm}/v_{dd}$) as function of the frequency. The OTA has presented a differential mode-gain of 29.66 dB and unity gain frequency of 21.88 MHz. The low-frequency common-mode and power supply rejection rates are 51.39 dB and 63.74 dB, and the common-mode and power supply gains remain under 0 dB in all the filter pass-band. The OTA post-layout simulation specifications at the V_{DD} of 0.4 V are shown in Table 5.4.

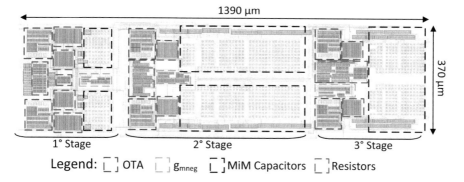

Fig. 5.8 Layout of the CxBPF circuit

5.1.4 CxBPF Measured Results

The complete CxBPF circuit was obtained by connecting all the OTAs, negative transconductors and passive devices presented in last subsections. The CxBPF complete layout is shown in Fig. 5.8. It has a size of $1390\,\mu\text{m} \times 370\,\mu\text{m}$ that results in a silicon area of $0.514\,\text{mm}^2$. A reasonable percentage of the total area is occupied by the programmable capacitor designed using the parallel association of some unitary MiM capacitors and by dummies devices. The active circuits, composed by the OTAs and the negative transconductors, occupy about 20% of the total silicon area.

In order to demonstrate the proposed CxBPF operation, we have designed and fabricated the circuit in a 180 nm six metal layers CMOS process. The microphotograph of this integrated circuit is shown in Fig. 5.9a. The fabricated die of $1.66\,\text{mm} \times 1.66\,\text{mm}$ was packaged using a CLCC package with 44 pins to perform all the measurements using a printed circuit board. Figure 5.9b shows the used CLCC 44 package with the cavity opened and closed.

The CxBPF circuit measurements were performed using the generic test board developed [6]. The equipment setup employed in the measurements process is shown in Fig. 5.10. A two-channel waveform generator is used to generate the I and Q signals in the frequency range of interest. The single-ended to differential mode conversion was performed using two transformer baluns at the input and one at the output. At the output, a High Z driver is used to match the output impedance with the impedance of 50 Ω of the spectrum analyzer. The output driver also has low input capacitance and a high input resistance in order not to degrade the CxBPF performance. All the output signals and the total equivalent output noise of the CxBPF were measured using the spectrum analyzer. The circuit was powered using symmetric ± 3 V batteries, and two voltage regulators were employed to obtain the +0.2 V and −0.2 V used in the CxBPF power supply. The batteries common node was used to generate the common-mode reference voltage of $V_{DD}/2$. The use of batteries instead of a standard power supply voltage was preferred to improve the

Fig. 5.9 Microphotograph of the fabricated IC in a 180 nm CMOS process (**a**), and the CLCC 44 package used to perform the measurements (**b**)

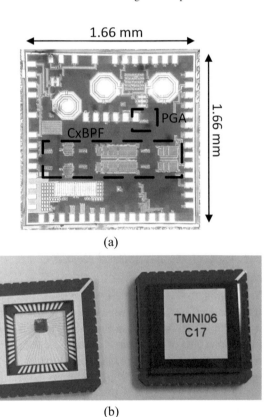

(a)

(b)

noise measurement accuracy. The digital configuration bits and the clock signal of the the serial digital input-interface were generated using an Arduino Uno R3 development board. It is connected to a personal computer using the USB interface where the circuit calibration can be performed. To transform the digital voltage level of the Arduino board from 5 V to 0.4 V a resistor-based logic level shifter was used. To ensure the external noise isolation, a custom made aluminum shield box was designed using some BNC type connector to perform the equipment connection.

The measurement process was performed first to calibrate the current references in order to adjust the negative transconductors and to present a 0 dB band-pass voltage gain. The programmable capacitors and resistors were also calibrated to set the filter bandwidth to 1 MHz and the IF to 2 MHz. Figure 5.11 shows the measured transfer functions of the CxBPF for the desired and image input signals. The CxBPF presents a band-pass voltage gain of 0 dB and a band-pass range from 1.5 MHz to 2.5 MHz, as designed. For the image signal the transfer function has −34 dB gain at 2 MHz, that results in an IRR of 34 dB. Both specifications are enough to satisfy the BLE standard requirement.

The integrated input-referred noise (IRN) was measured according to the output noise density. As shown in Fig. 5.12, the CxBPF has an average output noise density

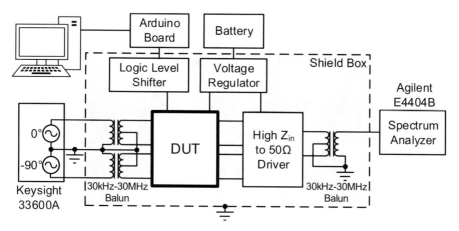

Fig. 5.10 The equipment setup used in the CxBPF measurements. Reprinted, with permission, from [6]

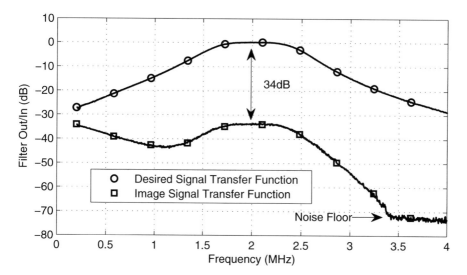

Fig. 5.11 Measured complex BPF transfer function for the desired and image signals. Reprinted, with permission, from [6]

of $180\,nV/\sqrt{Hz}$ at the pass-band that results in an IRN of $216\,\mu V$. As the low energy RF receiver topology is indeed to have a low noise amplifier (LNA) in the front-end part, the obtained IRN value does not affect the receiver sensitivity.

The CxBPF out of band input third-order intercept point (IIP3) was measured using two tones at 4 MHz and 6 MHz, respectively. Figure 5.13 shows the measured output power versus the input power for the fundamental and the third-order intermodulation (IM3) at 4 MHz and 2 MHz, presenting an IIP3 of 1.53 dBm.

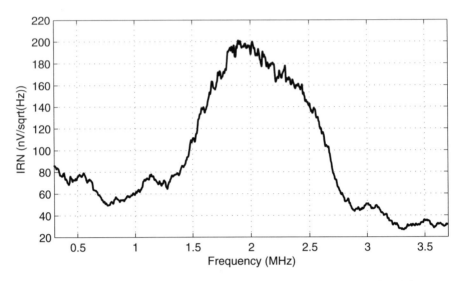

Fig. 5.12 Measured equivalent output noise density. Reprinted, with permission, from [6]

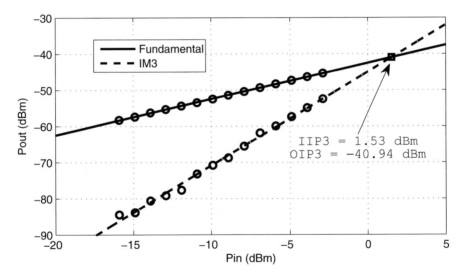

Fig. 5.13 Measured third-order interception point (IIP3). Reprinted, with permission, from [6]

Figure 5.14 presents the measurement of the spurious-free dynamic range (SFDR) for a complex input signal of −25 dbm of power and IF of 2 MHz. The third harmonic presents the highest spurious at 6 MHz with −77.7 dbm of power. It results in an SFDR of 52.7 dBc.

The total current drained from the power supply was 164 μA that results in a total power dissipation of 65.6 μW at 0.4 V or 10.9 μW per pole. The rest of the measured specifications and some results of other Bluetooth filters from the

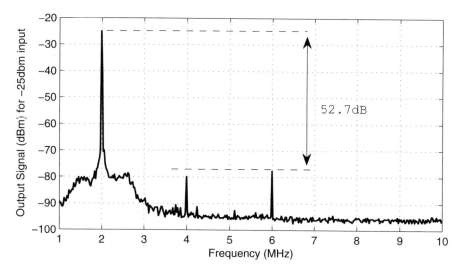

Fig. 5.14 Measured spurious-free dynamic range (SFDR)

Table 5.5 CxBPF measured specifications and comparison with other Bluetooth filters

Specification	This work	TCASII'17 [17]	TCASI'13 [1]	JSSC'10 [3]	Unit
Technology	180	180	180	90	nm
Voltage	0.4	1.8	1.8	0.6	V
Type	RC	RC	CA-RC	RC	–
	CxBPF	LPF	CxBPF	CxBPF	–
Order	3	4	4	6	–
Bandwidth	1.0	0.6	1.0	1.0	MHz
fc	2.0	–	3.0	1.0	MHz
Power	65.6	500	1000	6000	μW
Power/pole	10.9	125	125	500	μW
Noise	216	105[a]	73	130	μV
Out-of-band IIP3	1.53	25	29	−2.0[b]	dBm
SFDR	52.7	65.6	65.8	36.2	dB
IRR	34.0	–	56.0	33.0	dB
Silicon Area	0.51	0.13	0.40	–	mm^2
Meas./Sim.	Meas.	Meas.	Meas.	Meas.	–
FoM (Eq. 5.5)	0.0127	0.109	0.0214	7.744	pJ

[a]Noise of [1] was estimated using the spectral noise density
[b]IIP3 of [3] is the receiver IIP3 for the minimum gain setting
Reprinted, with permission, from [6]

literature are shown in Table 5.5. These works were compared by using the Figure of Merit (FoM) given in [1], expressed by Eq. 5.5.

$$FoM = \frac{Power}{N_{poles}.SFDR.f_{cutoff}} \tag{5.5}$$

Where: $Power$ is the filter power dissipation, N_{poles} is the number of poles, SFDR is the filter spurious-free dynamic range, and f_{cutoff} is the cutoff frequency in LPF and the center frequency in BPF.

This work has presented comparable specifications values, the best FoM, and the smallest power dissipation among the Bluetooth filters, even operating with a power supply of only 0.4 V. Further, we can compare the FoM value related to the supply voltage and to the power per pole of the Bluetooth publications compared in Table 5.5 and some state-of-the-art active filters for other applications [7, 8, 11, 15, 21, 22]. Figure 5.15 shows the FoM comparisons where we can conclude that our CxBPF circuit has also shown the best FoM among the state-of-the-art works, besides presenting the smallest operation voltage and power per pole.

5.2   Programmable Gain Amplifier

In this section an ultra-low voltage (ULV) and ultra-low power (ULP) programmable gain amplifier (PGA) using a closed-loop single-stage operational transconductance amplifier (OTA) suitable for low energy direct-conversion RF receivers is proposed.

The schematic of the proposed PGA is shown in Fig. 5.16. It is composed of a single-stage OTA, feedback resistors (R), programmable input resistors (R_V) and a programmable input negative transconductor ($g_{mneg}V$). As presented in Sect. 3.1, the compensation of the OTA low voltage gain and the resistive load sensitivity can be performed using an input negative transconductor. The PGA voltage gain of R/R_V is obtained when $g_{mneg}V$ is equal to $-(1/R_V + 1/R)$. The most challenge of the proposed circuit is ensure that $g_{mneg}V$ is changed to $-(1/R_V + 1/R)$ for different values of R_V to compensate the closed-loop in all the voltage gain range.

As shown in Fig. 5.16, the voltage gain is programmed with three thermometer coded bits (S_0-S_2) that open or close the switches and change the equivalent value of R_V and $g_{mneg}V$. The design of R_V is performed using multiples values of R in order to obtain a 6 dB gain step. The $g_{mneg}V$ is designed using a parallel association of four multiples transconductances of g_{mneg} and, choosing $g_{mneg} = -1/R$, the optimal value for $g_{mneg}V$ is obtained for any value of R_V. We have used R equal to 100 kΩ and, consequently, the g_{mneg} should be equal to 10 μS.

The proposed circuit was designed and fabricated in the TSMC 180 nm CMOS process. The design was performed to operate with the power supply of 0.36 V, which is only 20% of the 1.8 V process nominal voltage. This voltage value was chosen to evaluate the operation at the lowest bandwidth of a BLE receiver.

The resistor R was implemented with the process high resistivity poly material, and eight 12.5 kΩ series resistors were used to obtain the resistance of 100 kΩ. The switches employed to implement the programmable input resistor were performed using native NMOS transistors in order to obtain a low switch series resistance. The

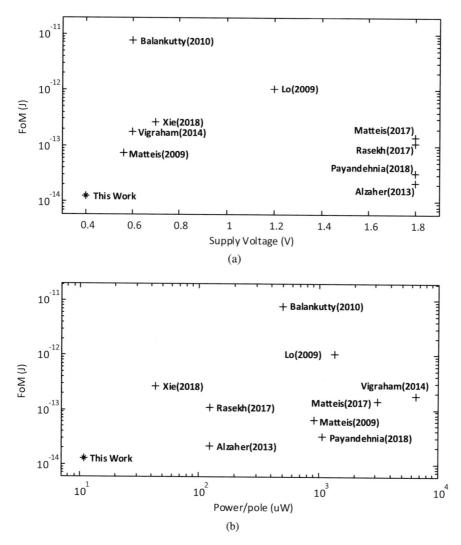

Fig. 5.15 Comparison of the FoM versus supply voltage (**a**), and the FoM versus the power per pole (**b**) of this work and other previously reported works

used transistors have the minimum channel length allowed to the native transistor of 0.5 μm, width of 5 μm and 5 multipliers. The rest of the circuit was implemented using Low-VT NMOS and PMOS transistors.

The following subsection presents some details of the OTA and the programmable negative transconductor implementation.

Fig. 5.16 Proposed PGA
using a programmable input
negative transconductance

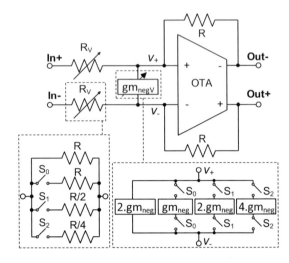

5.2.1 OTA Implementation

The OTA applied in the PGA implementation is the same circuit used in the CxBPF, presented in the last section. Its specifications were shown in Table 5.4 for a power supply of 0.4 V. However, due to the OTA auto compensation, it works well down to 0.3 V, but the bandwidth is insufficient at this voltage. In this circuit, we have used as reference the voltage of 0.36 V that present a unity gain frequency approximately equal to 3 MHz, and the OTA has a power dissipation around 7 μW.

5.2.2 Programmable Negative Input Transconductor

The main challenge of the PGA implementation is to preserve the match between the negative transconductor and the resistor to obtain the proper compensation. Thus, the negative transconductor should have a smaller sensitivity to the PVT variations and should have a trimming capability to adjust its value after the fabrication according to the resistor process variations.

The schematic of the proposed negative transconductor is shown in Fig. 5.17. It is composed of the proposed negative g_m cell, presented in Sect. 3.2, and three extra negative g_m cells that can be turned on or off. The four negative g_m cells were designed using the same transistor sizes but with different multiplicity factors and sharing the same bias circuits. The negative transconductance value of each cell is defined by the multiplicity factor, as presented in Fig. 5.16. The use of the same constant g_m bias for all the negative g_m cells makes the circuit more efficient and smaller. For the replica bias, the same circuit is used to bias all the PMOS bulk to save power. Furthermore, as the error amplifier is connected to all the PMOS bulks,

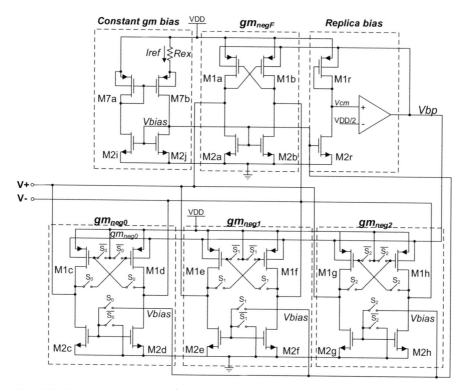

Fig. 5.17 Proposed programmable negative input transconductor

the loop stability is guaranteed due to the high equivalent bulk to ground parasite capacitance.

The programmability was performed by turning some of the negative g_m cells on or off and keeping all the g_m cells directly connected to the input nodes. This strategy was used instead of using series switches, as shown in Fig. 5.16, because of the high series resistances of the ULV switches that affect the equivalent negative transconductance value. Additionally, by turning the g_m cells on or off is possible to save power when in the low gain mode. To turn the negative g_m cell off the PMOS gate is connected to V_{DD} while the NMOS gate is connected to ground. In the on-mode the PMOS gate is connected to the **V+** and **V−** nodes and the NMOS gates are connected to V_{bias}.

The g_{mneg} was designed to be equal to 10 μS in order to perform the compensation when R is equal to 100 kΩ. The CMOS transistors were carefully sized to reduce the noise contribution at the OTA inputs and to reduce the mismatch effects. All the transistors were considered to have a channel length of 1 μm, as considered in the CxBPF circuit implementation. Table 5.6 show all the transistor channel widths, multipliers and the external resistor used to the constant g_m bias.

Table 5.6 Parameters values used in the PGA negative transconductors implementation

Parameter	W	M_{ab}	M_{cd}	M_{ef}	M_{gh}	M_{ij}	M_r
$M1x$	1.71 μm	2	1	2	4	1	1
$M2x$	8.32 μm	2	1	2	4	1	1
$M3x$	9.42 μm	1					
$M4x$	3.76 μm	1					
Parameter	W	M_a	M_b				
$M7x$	1.05 μm	5	14				
Other parameters	Value						
$Rext$	36.3 kΩ						

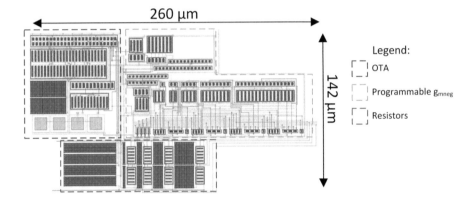

Fig. 5.18 Layout of the PGA circuit

5.2.3 *Measured Results*

The layout of the PGA was designed using the Cadence EDA tool and the circuit occupies an actual silicon area of 0.0243 mm², as shown in Fig. 5.18. The PGA was fabricated and the circuit specifications were measured using the same equipment setup used in the CxBPF characterization.

The programmable negative transconductor was first calibrated by adjusting the reference current using the external resistor. Matching the equivalent $g_{m_{negV}}$ to the actual value of $-(1/R_V + 1/R)$ is required because of the resistor process variation. Due to the resistor mismatch, the voltage gain could not be wholly compensated to obtain the exact gain value and 6 dB step. Additionally, the PGA voltage gain and bandwidth are related to the negative input transconductance. Figure 5.19 presents the measured results of the differential-mode gain and the bandwidth obtained with the variation of I_{ref} from 0.5 μA to 2.6 μA during the "111" gain mode calibration. The bandwidth is inversely related to I_{ref} whereas the gain increases for I_{ref} from 0.5 to 1.6 μA and decreases for $I_{ref} > 1.6$ μA due to the limitation in the replica

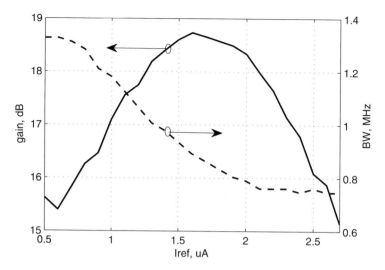

Fig. 5.19 The PGA gain and bandwidth variation as a function of the I_{ref} bias

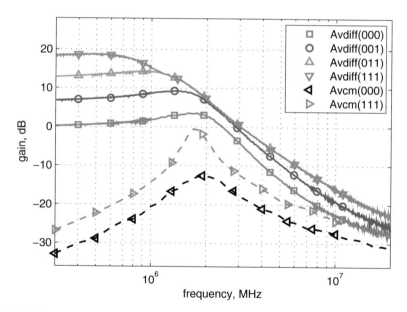

Fig. 5.20 Measured transfer functions of the PGA differential-mode (Av_{diff}) and common-mode (Av_{cm}) gains

bias voltage from 0 to 0.36 V. In this design, we used $I_{ref} = 1.4$ μA to obtain the gain and bandwidth of 18.4 dB and 0.98 MHz, respectively.

Figure 5.20 shows the PGA measured transfer function for the differential-mode (Av_{diff}), for all the gain modes, and the common-mode (Av_{cm}) gains,

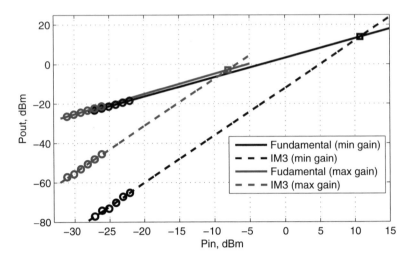

Fig. 5.21 Measured input third-order inter-modulation intercept point (IIP3)

for the minimum and maximum gain modes. Av_{diff} presents a programmability range from 0.2 dB to 18.4 dB with a step of approximately 6 dB. The PGA bandwidth has the highest value of 2.85 MHz at the 0 dB gain mode ("000") and the lowest bandwidth of 0.98 MHz at the 18 dB gain mode ("111"). The Av_{diff} transfer function presents small peaks in the passband because no external feedback capacitor was applied in this implementation, as analyzed in Sect. 3.1.1. Av_{cm} is dependent on the gain mode, and it is always lower than 0 dB. The CMRR, defined as Av_{diff}/Av_{cm}, is higher than 20 dB in the whole 0.98 MHz bandwidth.

Figure 5.21 shows the PGA out off band third-order intercept point measurements. It was measured by using two tones at the frequencies of 2.2 MHz and 4.2 MHz that results in the third-order intermodulation product (IM$_3$) at 200 kHz. The PGA presents an IIP$_3$ from −8.06 to 10.78 dBm, from the maximum to the minimum gain mode.

The measured DC power dissipation of the PGA at V_{DD} of 0.36 V is from 8.9 μW to 15.4 μW, depending on the gain mode.

Table 5.7 shows the rest of the measured specifications and a comparison with two low-power PGAs from the literature. The results present comparable specification values, and our PGA has obtained the smallest power dissipation, besides being able to work with a third of the supply voltage. The voltage gain and the input referred noise can be improved using multiples cascaded PGAs. Additionally, a capacitor can be added in parallel with the feedback resistor to also work as a channel selection filter for direct-conversion low-energy RF receivers, as used in [13], and to reduce the differential-mode peak in the passband.

Table 5.7 Measured results and comparison with other works from the literature

Specifications	This work				References		Unit
Gain mode	000	001	011	111	[24]	[14]	–
Power	8.9	9.4	11.2	15.4	56	55	μW
Diff. gain	0.2	6.7	12.8	18.4	−14/33	4/55	dB
Bandwidth	2.85	2.36	1.71	0.98	5.0	0.54	MHz
CMRR @300kHz	32.7	35.2	39.3	45.3	–	–	dB
PSRR @300kHz	16.9	21.5	25.2	27.4	–	–	dB
Input. ref. noise density	246	248	269	194	45	16.7	nV/$\sqrt{Hz}$
Vin_{pp} for THD=1%	266	189	63.2	26.5	–	–	mV
IIP3	10.78	4.51	−1.52	-8.06	–	–	dBm
Supply voltage	0.36				1.0	1.0	V
Technology	180				90	130	nm
Actual silicon area	0.0243				0.16	0.06	mm^2
Capacitive load	4				–	–	pF

5.3   Second-Order Low-Pass Filter with Integrated Programmable Gain Amplifier

Based on the previous circuits we have proposed a second-order active filter with integrated programmable gain capability. It can be applied in the baseband section of direct-conversion low energy receivers to select the desired channel, to reject the adjacent and alternate channels and to amplify the received signal.

The proposed circuit is based on the active-RC Tow-Thomas topology using two single-stage OTAs and two input-negative transconductors, as shown in the schematic of Fig. 5.22. The programmable resistor R1 and the programmable transconductor g_{mneg1} are employed to change the voltage gain of the filter, while g_{mneg2} is used to compensate the filter second-stage loop gain. The use of the programmable transconductor at the input of the first OTA has also increased the OTA bandwidth and keeps the OTA gain even with low values of R1.

In this circuit, the same variable transconductor used to implement the PGA circuit of Sect. 5.2 was used, as shown in Fig. 5.17. However, we have added two new transconductors to work with five bits to improve the programmability. Additionally, we have applied the OTA with the novel individual bulk control, as presented in Sect. 3.3.2.

The following subsections show the circuit design, its implementation using the Global Foundries 8HP 130 nm BiCMOS process and some post-layout simulation results.

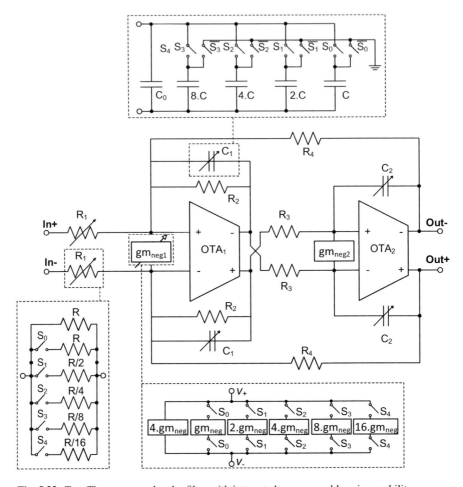

Fig. 5.22 Tow-Thomas second-order filter with integrated programmable gain capability

5.3.1 Filter Design

The filter design was based on the requirements of cutoff frequency and quality factor (Q_{filter}). In a BLE direct-conversion receiver the low-pass filter, placed after the down-conversion mixer, should select the desired information in a bandwidth of 1 MHz. Thus, the low pass filter should have a cutoff frequency higher than 500 kHz. A common choice is to design the filter with 600 kHz of bandwidth in order to avoid the 3 dB attenuation at the channel corners [14, 17].

The quality factor and the complex conjugate poles angular frequency (ω_0) of the Tow-Thomas active-RC filter can be estimated using Eqs. 5.6 and 5.7 [17].

$$Q_{filter} = \sqrt{\frac{R_2^2}{R_3.R_4} \cdot \frac{C_1}{C_2}} \tag{5.6}$$

$$\omega_0 = \frac{1}{\sqrt{R_3.R_4.C_1.C_2}} \tag{5.7}$$

We have chosen the Q_{filter} of $1/\sqrt{2}$ to present a Butterworth behavior. The match between the resistors and the negative transconductor was preserved by choosing $R_3 = R_4 = R$ and $R_2 = R/2$, where R was defined to be equal to 100 $k\Omega$. Thus, the unity transconductance cell used to implement the variable transconductor was defined to be approximately equal to $1/R$, to reach almost $10\mu S$. The constant negative transconductor employed at the OTA2 input was defined to be 8.5 μS, resulting in a integrator safety margin of 15%. To obtain $Q_{filter} = 1/\sqrt{2}$ and $\omega_0 = 600 \times 10^3/2\pi$ rad/s we used $C_1 = 2.C_2 = 3.75$ pF. Assuming the full compensation of the single-stage OTA low voltage gain and the resistive load effect, the low-frequency PGA voltage gain is equal to R_4/R_1. In order to find a programmable gain from 0 dB to 30 dB with 6 dB step, the resistor R1 should be programmable from 100 $k\Omega$ to 3.125 $k\Omega$ using five thermometer control bits.

As the integrated metal-insulator-metal (MiM) capacitors and the high resistivity P+ Poly resistors have variations of $\pm 10\%$ and $\pm 15\%$, respectively, C_1 and C_2 were designed as programmable capacitors to allow the tune on ω_0 and Q_{filter} after the fabrication. Table 5.8 shows all the parameters values used to implemented the Tow-Thomas LPF with integrated programmable-gain capability.

The circuit operation is very dependent on the switches used in the programmable resistors and capacitors. The design of the switches used to implement the programmable resistor R_1 is critical due to the match needed between R_1 and g_{mneg1}. If the switch mode-on series resistance is too high the equivalent resistor association is higher than the target value required to match with g_{mneg1}, generating a gain compensation error and reducing the stability margin.

Table 5.8 Parameters used to implement the Tow-Thomas LPF with integrated programmable-gain capability

Parameter	Value	$n°$ of bits
R_1	3.125 to 100 kΩ	5
R_2	50 kΩ	–
R_3	100 kΩ	–
R_4	100 kΩ	–
C_1	2 to 5 pF	4
C_2	$C_1/2$	4
R	100 kΩ	–
C	0.2 pF	–
C_0	2 pF	–
g_{mneg1}	40 to 350 μS	5
g_{mneg2}	8.5 μS	–
g_{mneg}	10 μS	–

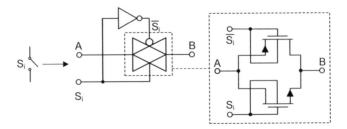

Fig. 5.23 The used gate connected to bulk transmission gate switch

Due to the reduced gate to source voltage to set the switches on at the ultra-low voltage operation, we have employed bulk connected to gate transmission gate switches to implement the most critical switches. Figure 5.23 shows the schematic of the transmission gate used, where the bulk terminal is connected to the gate terminal in both PMOS and NMOS transistors. The use of the bulk voltage reduces the transistor threshold voltage in the on-mode, reducing the switch on-resistance and increasing the ratio between the on-mode and off-mode switch resistance. The switches implementation were performed using both L=0.33 μm, Wp=14 μm and Wn=5 μm. In some switches, the transistor multiplicity factor was increased to further reduce the series resistance.

5.3.2   Negative Transconductors Implementation

As mentioned, we have added two new transconductors to the same variable transconductor used to implement the PGA circuit of Sect. 5.2 to work with five control bits.

We have designed a negative transconductor of 10 μS and used multiples of that to obtain the $4.g_{mneg}$, $8.g_{mneg}$ and the $16.g_{mneg}$. The 8.5 μS negative transconductor was designed using a channel width 15% lower than the values used in the 10 μS circuit implementation. The same replica bias and constant g_m bias circuits were applied in all the negative transconductors. Table 5.9 shows the devices sizes used to implement the negative transconductors. The same devices name of the circuit shown in Fig. 5.17 were used in Table 5.9.

The phase margin of the replica bias loop was improved by reducing the error amplifier bandwidth by connecting all the PMOS transistors bulk terminal to it output. Additionally, it bandwidth was further reduced using a low current bias and higher values of channel length for the M3a/b and M4a/b. The higher L employed also mitigate the effects of the mismatch in the error amplifier.

The layout of the negative transconductor was designed with the Cadence EDA tool, and is shown in Fig. 5.24. The circuit occupies an area of 0.017 mm^2, including the programmable negative transconductor connected to OTA1 and the constant negative transconductor connected to OTA2.

Table 5.9 Device parameters used to implement a 10 μS negative transconductor

Parameter	W	L	M
$M1a = M1b$	7.23 μm	1.50 μm	1
$M1r$	7.23 μm	1.50 μm	2
$M2a = M2b = M2c = M2d$	2.07 μm	1.50 μm	1
$M2r$	2.07 μm	1.50 μm	1
$M7a$	2.20 μm	1.00 μm	7
$M7b$	2.20 μm	1.00 μm	12
$M3a = M3b$	7.97 μm	5.00 μm	1
$M4a = M4b$	3.85 μm	20.00 μm	1
Other parameters	Value		
R_{ex}	39.3 kΩ		

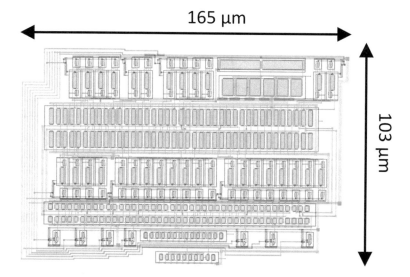

165 μm

103 μm

Fig. 5.24 Layout of the programmable and constant value negative transconductor

5.3.3 OTA Implementation

The OTA implementation is based on the circuit proposed in Sect. 3.3.2, that includes both the output common-mode voltage and current compensation. The OTA was designed to operate at the V_{DD} of 0.4 V and to present a GBW compatible with the LPF circuit. As suggested in [23], the GBW of the single-stage OTAs should be higher than $8 . Q_{filter} \cdot f_{cutoff}$. Considering the cutoff frequency and Q_{filter} used in the LPF, the GBW should be higher than 3.4 MHz. However, as the LPF will also be applied as programmable-gain amplifier, we designed it to present an unity gain frequency of 13.8 MHz.

The circuit was designed using standard-V_T transistors, and all the transistor sizes and the values of the resistors and capacitors used in the OTA implementation

Table 5.10 Device parameters used in the OTA implementation

Device	W	L	M
$M4a = M4b$	15.67 μm	1.00 μm	4
$M4r$	15.67 μm	1.00 μm	2
$M5a = M5b$	1.97 μm	0.50 μm	10
$M5c = M5d$	1.97 μm	0.50 μm	1
$M5r$	1.97 μm	0.50 μm	1
$M6a = M6b$	8.05 μm	0.50 μm	10
$M6c = M6d$	8.05 μm	0.50 μm	1
$M7a$	1.69 μm	5.00 μm	7
$M7b$	1.69 μm	5.00 μm	11
$M8a = M8b$	8.05 μm	0.50 μm	2
$M9a = M9b$	1.97 μm	0.50 μm	2
$M10a = M10b$	7.97 μm	5.00 μm	3
$M11a = M11b$	3.85 μm	20.00 μm	1
Cc (MOS CAP)	10.00 μm	10.00 μm	14 fingers
Device	Value		
Rex	10kΩ		
$Rcma = Rcmb$	100 kΩ		
$Ccma = Ccmb$	0.5 pF		

are shown in Table 5.10. The loop of the NMOS bulk bias control was stabilized by using a large Cc capacitor placed at the ErrAmp2 output. To save silicon area it was implemented using the gate capacitance of a multi-finger NMOS transistor with drain and source connected to GND.

The OTA1 and OTA2 used in the Tow-Thomas LPF implementation were designed to be equal in order to share the same NMOS bulk bias control and to reduce the power dissipation. To reduce the mismatch and the silicon area we designed the OTA1 and OTA2 layout together in a single layout block, as shown in Fig. 5.25. The layout has the size of 156 μm × 120 μm = 0.019 mm^2 or 0.0095 mm^2 per OTA.

Table 5.11 shows the OTA specifications obtained with post-layout simulations, considering an output capacitance load of 5 pF. Figure 5.26 shows the open-loop post-layout simulation for the differential-mode, common-mode and power-supply gains as function of the frequency. The low-frequency gain is 26.5 dB and the common-mode and power-supply gains are lower than 0 dB up to 2 MHz. In lower frequencies, the common-mode rejection rate (CMRR) and the power supply rejection ratio (PSRR) are 40.19 dB and 48.76 dB, respectively. Figure 5.27 shows the closed-loop transient simulation for a pulse input signal. The slew-rate obtained is approximately 9.4 V/μs in both the rising and falling. The total current drained from the 0.4 V power supply by OTA1 and OTA2 is 57.5 μA, which results in an average power dissipation of 11.5 μW per OTA.

In order to analyze the improvements of the proposed OTA, we have performed Monte Carlo simulations with 1000 samples, including process and mismatch analysis. Table 5.12 shows the average and the standard deviation values for some

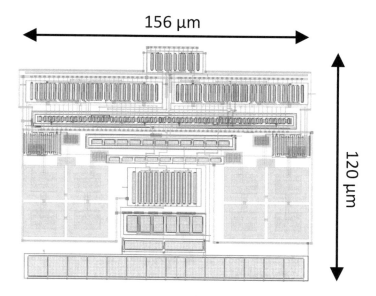

Fig. 5.25 Layout of the OTA1 and OTA2 amplifier used in the filter implementation

Table 5.11 Post-layout
simulation results of the 0.4
V OTA

Specifications	Value
Technology	130 nm
Supply voltage	0.4 V
Differential-mode gain	26.50 dB
Unity gain frequency	13.79 MHz
Common-mode gain	−13.69 dB
CMRR	40.19 dB
Power supply gain	−22.26 dB
PSRR	48.76 dB
Slew Rate	9.38 V/μs
Power dissipation	11.50 μW
Area	0.0095 mm^2
Capacitive load	5 pF

specifications in the circuit with and without the proposed NMOS bulk control. The simulations without the NMOS bulk control were performed with the NMOS bulk tied to $V_{DD}/2$ voltage. The low-frequency gain (Av_o) and the DC output common-mode voltage are not so affected by the NMOS control because it is mostly controlled by CMFB feedback. The standard deviation of the current drained by each CMOS inverter in the main OTA was reduced from 3.63 μA to 1.07 μA by using the NMOS bulk control. As a consequence of this, the standard deviation of the GBW was reduced from 3.25 MHz to 1.25 MHz. The average of the total current drained from the VDD (OTA 1 + NMOS bias control) was increased from 26.83 μA to 32.16 μA due to the power dissipation of the NMOS bias control

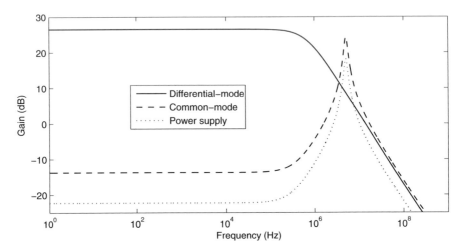

Fig. 5.26 Post-layout simulation results of the differential-mode, the common-mode and the power supply gains for the proposed OTA

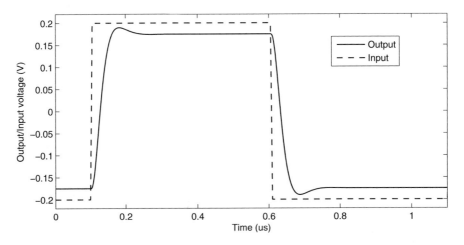

Fig. 5.27 Post-layout simulation pulse response for the proposed OTA

circuit. However, the standard deviation was reduced from 7.30 μA to 3.39 μA. It has a smaller reduction factor in comparison to the other specifications because the NMOS bias is not applied in all the OTA NMOS transistors. Figure 5.28a–c shows the histograms obtained with the Monte Carlo simulation. The histograms were generated considering the same axis range and the number of bars. By analyzing this histograms we can graphically verify the reduction in the variability provided by the use of the proposed NMOS bulk control circuit.

Table 5.12 Comparison of the process variability in some specifications of the designed OTA with and without using the proposed NMOS bulk control. The results were based on the Monte Carlo process and mismatch simulation with 1000 samples

Specifications	Without NMOS control Avg/std	With NMOS control Avg/std
Av_{00}	26.18/0.75 dB	26.38/0.59 dB
DC output CM voltage	199.73/4.98 mV	199.90/4.56 mV
Inverter current	10.63/3.63 μA	10.07/1.07 μA
GBW	13.62/3.25 MHz	13.26/1.25 MHz
Total Current	26.83/7.30 μA[a]	32.16/3.39 μA

[a] The current drained by the NMOS bulk control circuit is 5.5 μA

5.3.4   Post-Layout Simulated Results of the Programmable-Gain LPF

The layout of the complete programmable gain LPF was designed using the Cadence® EDA tools. As can be seen in Fig. 5.29 the layout occupies a silicon area of 0.0973 mm² (345 μm to 282 μm), excluding the area of the I/O PADs and the register bank. The layout extraction was performed and the circuit specifications were simulated by using some test benches and considering a capacitive load of 5 pF.

Figure 5.30 shows the transfer function of the LPF for all the voltage gain modes from 0 dB to 30 dB. The cutoff frequency changes a little according to the gain mode. It can be compensated by adjusting the programmable capacitors C1 and C2 to present 600 kHz in all the gain modes. In the highest gain mode of 30 dB, the quality factor is higher than $1/\sqrt{2}$ as can be seen in Fig. 5.30, where the transfer function has a peak near to the cutoff frequency. The changing in the quality factor occurs due to the increase in the negative input transconductance at the input of OTA1, and it is very depended on the matching between the equivalent negative transconductance and the equivalent resistance. As the R1 resistor is programmable, the switch series resistance tends to increase this effects. As presented in the previous section, these switches were optimized to reduce the series resistance and, consequently also reduce the effect on the quality factor.

The filter common-mode rejection rate (CMRR) is higher than the OTA CMRR because of the extra common-mode attenuation provided by the input negative transconductor. Figure 5.31a shows the transfer function of the filter CMRR. At low frequency, it is in the range of 50 to 70 dB, according to the gain mode. The higher the gain, the higher the CMRR is. The CMRR remains over 0dB up to the frequency of 3 MHz, about five times the filter cutoff frequency.

Differently, the power supply rejection rate (PSRR) is not improved by the negative input transconductance. It is very similar to the PSRR of the OTA but has a lower value for reduced closed loop gain. The transfer function of the filter PSRR is shown in Fig. 5.31b for all the gain modes. At lower frequencies, the PSRR is

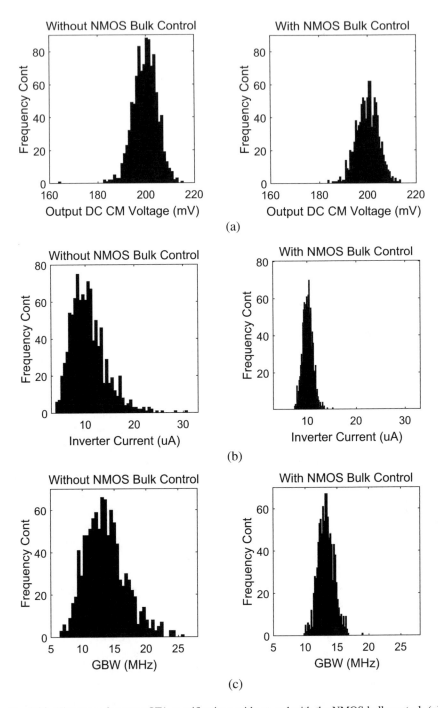

Fig. 5.28 Histogram for some OTA specifications without and with the NMOS bulk control: (**a**) Output DC common-mode Voltage, (**b**) CMOS inverter current, and (**c**) GBW

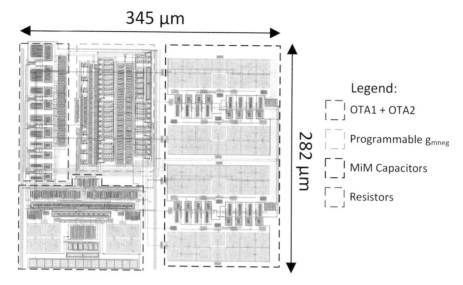

Fig. 5.29 Complete layout of the Tow-Thomas LPF with integrated programmable gain

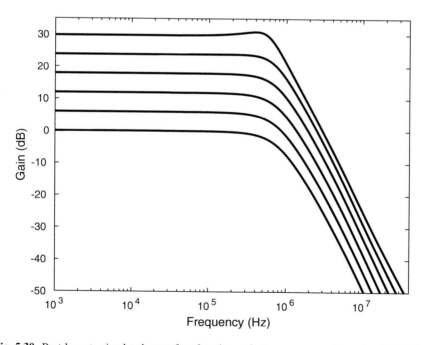

Fig. 5.30 Post-layout simulated transfer function of the programmable-gain LPF. The thermometric-coded control bits were changed to set the desired voltage gain from 0 dB to 30 dB with 6 dB step

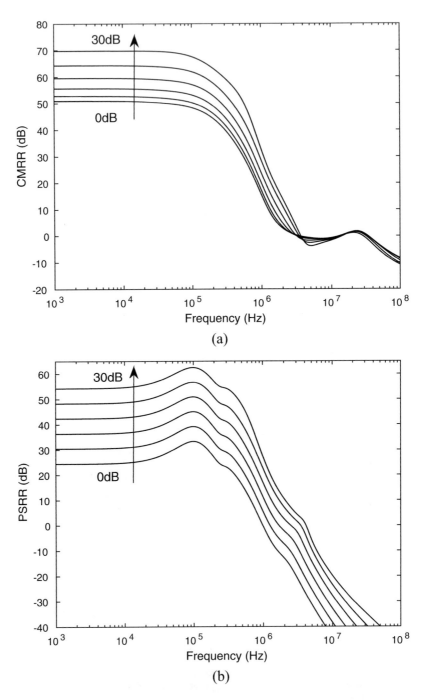

Fig. 5.31 Post-layout simulated common-mode and power supply rejection rates of the programmable gain LPF as a function of frequency and the gain modes from 0 dB to 30 dB with 6 dB step: (**a**) CMRR and (**b**) PSRR

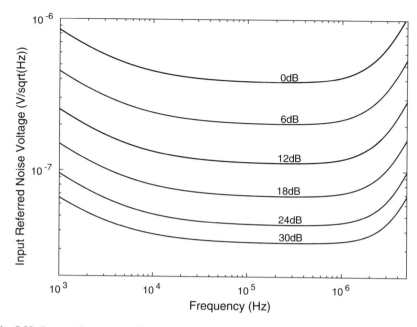

Fig. 5.32 Input-referred noise (IRN) voltage of the programmable-gain LPF as a function of the frequency for all the voltage gain modes

kept in the range from 24 dB to 54 dB. For all the gain modes it is over 0 dB for frequencies up to 1 MHz.

The circuits proposed in this work, using the input negative transconductor, have the output noise very dependent on the noise generated at the input of the OTA. The transistors sizes were optimized to reduce the input negative transconductance noise contribution. Figure 5.32 shows the frequency response of the input referred noise (IRN) density for all the gain modes. As expected, the higher the voltage gain, the lower the IRN is. The circuit has a minimum and a maximum IRN of 31.15 nV/$\sqrt{Hz}$ and 456.2 nV/$\sqrt{Hz}$ at the frequency of 100 kHz.

The filter dynamic range was evaluated by using the spurious-free dynamic range (SFDR) and the total harmonic distortion (THD) analysis. The values of SFDR and THD are dependent on the output voltage level. Figure 5.33a,b show the results of the post-layout simulation of SFDR and THD when the differential output voltage is changed from 10 mV to 800 mV. From 10 mV to 300 mV the maximum value of SFDR and the minimum value of THD are kept approximately equal to 55 dB and 0.2% for all the gain modes. From 300 mV of the differential output voltage, the SFDR and the THD start to reduce and increase, respectively. The maximum differential output swing expected to the LFP is 400 mV, and at this level, the SFDR remains over 50 dB, and the THD remains lower than 0.3%. Based on the output voltage limit of 400 mV, the maximum differential input voltage range should be from 12.5 to 400 mV according to the filter voltage gain.

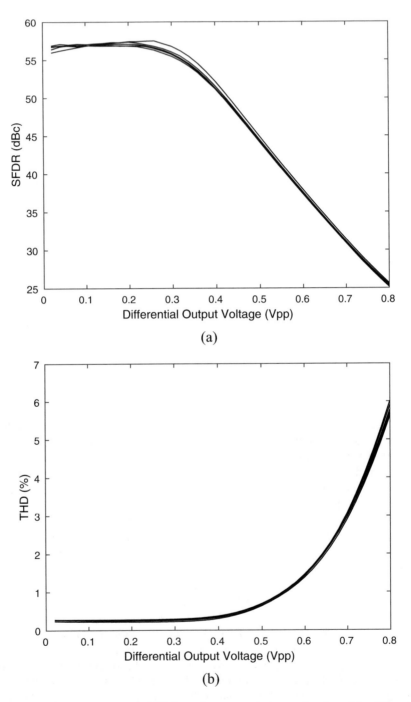

Fig. 5.33 The programmable gain LPF dynamic range analysis as a function of the differential output voltage in all the gain modes: (**a**) spurious-free dynamic range (SFDR) and (**b**) total harmonic distortion (THD)

Table 5.13 LPF post-layout simulation results

Specifications	Gain mode						Unit
	0dB	6dB	12dB	18dB	24dB	30dB	–
Gain	0.00	5.99	11.95	17.94	23.83	29.79	dB
Cutoff	619	622	627	644	680	716	kHz
Range of fc	0.41–1.1	0.41–1.1	0.42–1.1	0.43–1.1	0.46–1.2	0.52–1.1	MHz
Power	25.98	26.39	27.20	28.84	32.17	39.05	μW
Power/pole	12.99	13.20	13.60	14.42	16.09	19.53	μW
IRN Density	456.2	242.2	133.9	80.47	51.68	31.15	$V/\sqrt{Hz}$
IIP_3	17.6	10.9	5.1	−0.9	−6.8	−13.2	dBm
SFDR	54.18	51.17	51.29	51.56	52.18	51.46	dBc
THD	0.22	0.25	0.27	0.27	0.28	0.28	%
CMRR	50.87	52.77	55.66	59.55	64.29	69.74	dB
PSRR	24.46	30.44	36.40	42.39	48.32	54.26	dB
Area	0.0973						mm^2

Table 5.14 Comparison with some active-RC LPFs from the literature

Specifications	This work	JSSC'05 [4]	JSSC'09 [7]	JSSC'14 [21]	TCAS'17 [17]	Unit
Technology	130	180	130	65	180	nm
V_{DD}	0.4	0.45	0.55	0.6	1.8	V
Power/pole	13⇔19.5	135	875	6550	125	μW
Area	0.097	1.0	0.43	0.38	0.140	mm^2
Order	2	5	4	4	4	–
Gain	0⇔30	0	0	0	10	dB
Cutoff	0.4⇔1.1	0.153	11.3	70.0	0.6	MHz
IRN	31⇔456	200	–	–	126	$\frac{nV}{\sqrt{Hz}}$
IIP_3	−13.2⇔17.6	–	13.0	–	25	dBm
DR	54.2⇔51.5[a]	55.2[b]	60.0[b]	58.0[a]	65.6[a]	dBc
FoM	0.05⇔0.075	0.978	0.077	0.178	0.109	pJ

[a,b] Dynamic range (DR) based on SFDR[a] and on THD[b]

The circuit power dissipation is dependent on the gain mode as a function of the number of negative transconductance cell that are enabled. It varies from 25.98 μW to 39.05 μW, resulting in the power dissipation per pole in the range from 12.99 μW to 19.53 μW. The rest of the post layout simulated LPF specifications are shown in Table 5.13.

To compare the results of the proposed Tow-Thomas LPF with integrated PGA, we used the same Figure of Merit (FoM) as presented by Eq. 5.5 in the analysis of the CxBPF of Sect. 5.1.4.

Table 5.14 presents a comparison of the designed LPF with other LPF presented in the literature. The designed LPF in this work has presented comparable specifications, the smallest operation voltage, less power per pole and FoM, besides

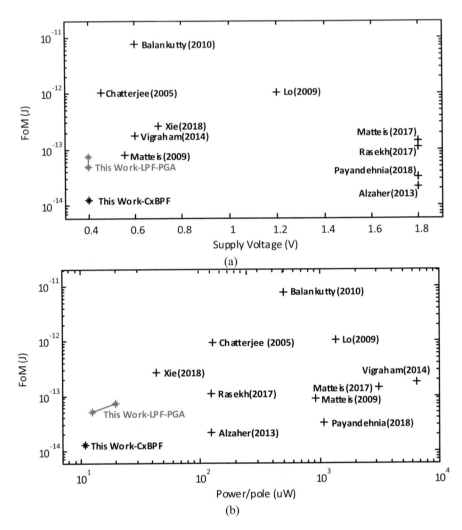

Fig. 5.34 Comparison of the FoM of this work and other previously reported work: (**a**) FoM versus supply voltage and (**b**) FoM versus the power per pole

performing a 30 dB range programmable gain capability. The power per pole is 6.4 times lower than the lowest power dissipation from the literature. To compare the FoM with other filters from the literature, we have repeated the graphs of the FoM versus the power per pole and the supply voltage, shown in Fig. 5.15, by including the results of the designed Two-Thomas LPF. These new graphs are shown in Fig. 5.34a,b, where is possible to see that the LPF has the smallest operation voltage and power dissipated per pole of all the compared works. It also has the best FoM among the low-pass filters compared. The programmable gain LPF filter design has a higher power dissipation per pole and higher FoM in comparison to the CxBPF presented in Sect. 5.1. These characteristics are related to the power dissipation added to implement the programmable negative transconductor and the OTA NMOS bulk control.

5.4 Conclusion

We presented in this chapter a 0.4 V ULP CxBPF, a 0.36 V PGA and a 0.4 V LPF with integrated PGA capability compatible with Low-IF and Zero-IF BLE RF receivers. All these circuits have employed the strategy presented in this book to reach the ULP operation by using the single-stage OTA and the negative input transconductor and the ULV operation was reached by using only two-stacked transistors and the bulk forward bias.

The PGA implementations used programmable negative input transconductors to match the input transconductance with the equivalent OTA input resistance at different gain modes. To our knowledge, it is the first time that this solution is presented in the literature.

The measured and post-layout performance showed comparable specifications, the best FoM and the smallest power dissipation among the compared works, including state-of-the-art papers. To our knowledge, our results reached the lowest power dissipation among the BLE BPF presented in the literature.

The designed circuits have some resistors and capacitor to be tunned after the fabrication. This is the main disadvantage of the active-RC circuits since the passive devices fabrication present increased variabilities. However, several strategies have been proposed in the literature to perform the automatic tune and calibration. Some of these techniques can be easily implemented at ULV and adapted to the proposed circuits [5, 10].

References

1. H.A. Alzaher, N. Tasadduq, F.S. Al-Ammari, Optimal low power complex filters. IEEE Trans. Circuits Syst. I Regular Papers **60**(4), 885–895 (2013). https://doi.org/10.1109/TCSI.2012.2209293
2. R.J. Baker, *CMOS: Circuit Design, Layout, and Simulation*, 3 edn. (Wiley, Hoboken, NJ, USA, 2011). https://doi.org/10.1002/9780470891179
3. A. Balankutty, S.A. Yu, Y. Feng, P.R. Kinget, A 0.6-V zero-IF/low-IF receiver with integrated fractional-N synthesizer for 2.4-GHz ISM-band applications. IEEE J. Solid State Circuits **45**(3), 538–553 (2010). https://doi.org/10.1109/JSSC.2009.2039827
4. S. Chatterjee, Y. Tsividis, P. Kinget, 0.5-V analog circuit techniques and their application in OTA and filter design. IEEE J. Solid State Circuits **40**(12), 2373–2387 (2005a). https://doi.org/10.1109/JSSC.2005.856280

5. S. Chatterjee, Y. Tsividis, P. Kinget, A 0.5V filter with PLL-based tuning in 0.18 um CMOS, in *ISSCC. 2005 IEEE International Digest of Technical Papers. Solid-State Circuits Conference, 2005*, San Francisco. (IEEE, 2005b), pp. 2004–2006. https://doi.org/10.1109/ISSCC.2005.1494091

6. L. Compassi-Severo, W. Van Noije, A 0.4-V 10.9-μ W/Pole third-order complex BPF for low energy RF receivers. IEEE Trans. Circuits Syst. I Regular Papers **66**(6), 2017–2026 (2019a). ISSN 15580806. https://doi.org/10.1109/TCSI.2019.2906206

7. M. De Matteis, S. D'Amico, A. Baschirotto, A 0.55 V 60 dB-DR fourth-order analog baseband filter. IEEE J. Solid State Circuits **44**(9), 2525–2534 (2009). https://doi.org/10.1109/JSSC.2009.2024801

8. M. De Matteis, A. Pipino, F. Resta, A. Pezzotta, S. D'Amico, A. Baschirotto, A 63-dB DR 22.5-MHz 21.5-dBm IIP3 fourth-order FLFB analog filter. IEEE J. Solid State Circuits **52**(7), 1977–1986 (2017). https://doi.org/10.1109/JSSC.2017.2693240

9. A.A. Emira, E. Sánchez-Sinencio, A pseudo differential complex filter for bluetooth with frequency tuning. IEEE Trans. Circuits Syst. II Analog Digit. Signal Process. (2003). https://doi.org/10.1109/TCSII.2003.818362

10. S. Kousai, M. Hamada, R. Ito, T. Itakura, A 19.7 MHz, fifth-order active-RC chebyshev LPF for draft IEEE802.11n with automatic quality-factor tuning scheme. IEEE J. Solid State Circuits **42**(11), 2326–2337 (2007). https://doi.org/10.1109/JSSC.2007.906189

11. T.Y. Lo, C.C. Hung, M. Ismail, A wide tuning range GmC filter for multi-mode CMOS direct-conversion wireless receivers. IEEE J. Solid State Circuits **44**(9), 2515–2524 (2009). https://doi.org/10.1109/JSSC.2009.2023154

12. K.W. Martin, Complex signal processing is not complex. IEEE Trans. Circuits Syst. I Regular Papers **51**(9), 1823–1836 (2004). https://doi.org/10.1109/TCSI.2004.834522

13. J. Masuch, M. Delgado-Restituto, *Ultra Low Power Transceiver for Wireless Body Area Networks*. (Springer International Publishing, Heidelberg, 2013a). https://doi.org/10.1007/978-3-319-00098-5

14. J. Masuch, M. Delgado-Restituto, A 1.1-mW-RX -81.4-dBm sensitivity CMOS transceiver for bluetooth low energy. IEEE Trans. Microwave Theory Tech. **61**(4), 1660–1673 (2013b). https://doi.org/10.1109/TMTT.2013.2247621

15. P. Payandehnia, H. Maghami, H. Mirzaie, M. Kareppagoudr, S. Dey, M. Tohidian, G.C. Temes. A 0.49-13.3 MHz tunable fourth-order LPF with complex poles achieving 28.7 dBm OIP3. IEEE Trans. Circuits Syst. I Regular Papers **65**(8), 1–12 (2018). https://doi.org/10.1109/TCSI.2017.2788466

16. A. Pipino, A. Liscidini, K. Wan, A. Baschirotto, Bluetooth low energy receiver system design. 2015 IEEE Int. Symp. Circuits Syst. (ISCAS) (1), 465–468 (2015). https://doi.org/10.1109/ISCAS.2015.7168671

17. A. Rasekh, M. Sharif Bakhtiar, Design of low-power low-area tunable active RC filters. IEEE Trans. Circuits Syst. II Express Briefs (2017). https://doi.org/10.1109/TCSII.2017.2658635

18. L.C. Severo, W.A.M.V. Noije, An optimization-based design methodology with PVT analysis for ultra-low voltage analog ICs, in *Conference on PhD Research in Microelectronics and Electronics (PRIME)* (IEEE, Lisbon, 2016), pp. 1–4

19. R.R. Silva, L.C. Severo, F. Sola, H.D. Hernandez, D.d. Santos, W.C. Aranda, W.A.M.V. Noije, A bluetooth low energy system analysis for low power applications, in *Workshop on Semiconductors and Micro & Nano Technology - Seminatec*, São Paulo (2017)

20. C. Upathamkuekool, A. Jiraseree-Amornkun, J. Mahattanakul, A low-voltage low-power complex active-RC filter employing single-stage opamp, in *2012 IEEE International Conference on Electron Devices and Solid State Circuit, EDSSC 2012*, pp. 2–5 (2012). https://doi.org/10.1109/EDSSC.2012.6482868

21. B. Vigraham, J. Kuppambatti, P.R. Kinget, Switched-mode operational amplifiers and their application to continuous-time filters in nanoscale CMOS. IEEE J. Solid State Circuits **49**(12), 2758–2772 (2014). https://doi.org/10.1109/JSSC.2014.2354641

22. Z. Xie, J. Wu, C. Chen, A Compact low-power biquad for active-RC complex filter. IEEE Trans. Circuits Syst. II Express Briefs **65**(6), 709–713 (2018). https://doi.org/10.1109/TCSII.2017.2784819
23. L. Ye, C. Shi, H. Liao, R. Huang, Y. Wang, Highly power-efficient active-RC filters with wide bandwidth-range using low-gain push-pull opamps. IEEE Trans. Circuits Syst. I Regular Papers **60**(1), 95–107 (2013)
24. C. Zhou, P. Harpe, S. Rampu, X. Wang, S. D'Amico, A. Baschirotto, K. Philips, G. Dolmans, H. De Groot, A 56μW VGA with 5MHz bandwidth and 47dB gain-range in 90nm CMOS. Proc. 2010 Int. Symp. VLSI Des. Autom. Test VLSI-DAT 2010 **1**(2), 91–94 (2010). https://doi.org/10.1109/VDAT.2010.5496699

Chapter 6
Conclusions

This book presented the development of active RC-filters and programmable gain amplifiers for BLE RF receivers with ULP dissipation and operating at the ULV range. The operation at the ULV range is important to obtain low-energy devices with improved lifetime. In the practical applications of IoT, the V_{DD} voltage used to supply the ULV circuits can be obtained using high efficient DC-DC converters when powered using batteries or energy harvesting circuits.

Additionally, the ULV operation can also be very useful in digital circuits operating at the minimum energy point (MEP), making easily the interface between the analog and digital domains [1, 2].

The key strategy used in this book to reach the ULP operation is based on using high-efficient inverter-based single-stage OTAs. The low voltage gain and loading effects, when in the closed-loop operation, were compensated by using an input connected negative transconductance. The analysis of the compensation technique considered the effects of the parasitic input and feedback capacitances and the equivalent output and input-referred noise. Based on these analyses, the optimal single-stage OTA compensation can be reached without instability issue and the noise power added by the negative transconductor can be estimated.

The strategy used to achieve the ULV operation was based on designing all the circuits using only two-stacked transistors, the bulk forward bias, and the proper transistor channel length design. The circuits were designed using fully-differential implementations to improve the dynamic range. An ULV negative transconductor using a replica circuit and the PMOS bulk forward bias was introduced in this work to reduce the variations on the input common-mode DC voltage and to extend the range of adjustable transconductance. The development of an improved ULV inverter-based OTA, combining a novel NMOS bulk replica bias with the common-mode feedback circuit, was also introduced in this work to reduce the variabilities on the output common-mode DC voltage and the current drained from the power supply without using any series transistor.

© The Author(s), under exclusive license to Springer Nature Switzerland AG 2022
L. C. Severo, W. A. M. Van Noije, *Ultra-low Voltage Low Power Active-RC Filters and Amplifiers for Low Energy RF Receivers*,
https://doi.org/10.1007/978-3-030-90103-5_6

In order to reduce the ULV circuit design effort, a design methodology based on the transistor operation point was also proposed in this book, and a computational tool was implemented. The proposed methodology was also added to the UCAF analog design tool [3] to improve the design space exploration efficiency on the design of ULV circuits.

The application of the proposed circuits was performed by designing active-RC filters and programmable gain amplifiers. A complex band-pass filter was designed and fabricated in a 180 nm CMOS process to operate at the IF of 2 MHz and 1 MHz of bandwidth. This circuit has presented 10.9 μW of power dissipation per pole, 52.7 dB of SFDR and 34 dB of image rejection rate when powered at 0.4 V. A programmable gain amplifier was fabricated on the same process to operate with only 0.36V of the power supply. The PGA presented power dissipation in the range from 8.9 to 15.4 μW, according to the gain mode from 0 to 18 dB, and the minimum bandwidth of 0.98 MHz. Based on the previous circuit a programmable gain Tow-Thomas low-pass filter was designed and fabricated in a 130 nm BiCMOS process. The post-simulated results shown programmable gain range from 0 to 30dB, power dissipation per pole from 12.99 to 19.53 μW, 54.18 dB of SFDR and CMRR over 50 dB when powered at 0.4 V. The designed circuits have presented the smallest operation voltage and power dissipation and the best figure of merit (FoM) when compared to other circuits present in the literature. The programmable negative transconductor used to implement the programmable gain amplifier and the Tow-Thomas biquad was introduced in this book.

References

1. M. Alioto, Ultra-low power VLSI circuit design demystified and explained: a tutorial. IEEE Trans. Circ. Syst. I: Regul. Pap. **59**(1), 3–29 (2012). https://doi.org/10.1109/TCSI.2011.2177004
2. N. Reynders, W. Dehaene, *Ultra-Low-Voltage Design of Energy-Efficient Digital Circuits*, 1st edn. (Springer, New York, 2015)
3. L.C. Severo, A. Girardi, A.B. de Oliveira, F.N. Kepler, M.C. Cera, Simulated annealing to improve analog integrated circuit design: trade-offs and implementation issues, in *Simulated Annealing - Single and Multiple Objective Problems* (Intech, London, 2012), pp. 261–283

Index

© The Author(s), under exclusive license to Springer Nature Switzerland AG 2022
L. C. Severo, W. A. M. Van Noije, *Ultra-low Voltage Low Power Active-RC Filters
and Amplifiers for Low Energy RF Receivers*,
https://doi.org/10.1007/978-3-030-90103-5

Printed in the United States
by Baker & Taylor Publisher Services